PLANNING, PROGRAMMING, BUDGETING, AND EXECUTION

IN COMPARATIVE ORGANIZATIONS

VOLUME 3

Case Studies of Selected
Non-DoD Federal
Agencies

MEGAN McKERNAN | STEPHANIE YOUNG | RYAN CONSAUL

MICHAEL SIMPSON | SARAH W. DENTON | ANTHONY VASSALO

WILLIAM SHELTON | DEVON HILL | RAPHAEL S. COHEN

JOHN P. GODGES | HEIDI PETERS | LAUREN SKRABALA

Prepared for the Commission on Planning, Programming, Budgeting, and Execution Reform

Approved for public release; distribution is unlimited

For more information on this publication, visit **www.rand.org/t/RRA2195-3**.

About RAND

The RAND Corporation is a research organization that develops solutions to public policy challenges to help make communities throughout the world safer and more secure, healthier and more prosperous. RAND is nonprofit, nonpartisan, and committed to the public interest. To learn more about RAND, visit www.rand.org.

Research Integrity

Our mission to help improve policy and decisionmaking through research and analysis is enabled through our core values of quality and objectivity and our unwavering commitment to the highest level of integrity and ethical behavior. To help ensure our research and analysis are rigorous, objective, and nonpartisan, we subject our research publications to a robust and exacting quality-assurance process; avoid both the appearance and reality of financial and other conflicts of interest through staff training, project screening, and a policy of mandatory disclosure; and pursue transparency in our research engagements through our commitment to the open publication of our research findings and recommendations, disclosure of the source of funding of published research, and policies to ensure intellectual independence. For more information, visit www.rand.org/about/principles.

RAND's publications do not necessarily reflect the opinions of its research clients and sponsors.

Published by the RAND Corporation, Santa Monica, Calif.
© 2024 RAND Corporation
RAND® is a registered trademark.

Library of Congress Cataloging-in-Publication Data is available for this publication.
ISBN: 978-1-9774-1238-6

Cover design by Peter Soriano; adimas/Adobe Stock images.

About This Report

The U.S. Department of Defense (DoD) Planning, Programming, Budgeting, and Execution (PPBE) process is a key enabler for DoD to fulfill its mission. But in light of a dynamic threat environment, increasingly capable adversaries, and rapid technological changes, there has been increasing concern that DoD's resource planning processes are too slow and inflexible to meet warfighter needs.[1] As a result, Congress mandated the formation of a legislative commission in Section 1004 of the National Defense Authorization Act for Fiscal Year 2022 to (1) examine the effectiveness of the PPBE process and adjacent DoD practices, particularly with respect to defense modernization; (2) consider potential alternatives to these processes and practices to maximize DoD's ability to respond in a timely manner to current and future threats; and (3) make legislative and policy recommendations to improve such processes and practices for the purposes of fielding the operational capabilities necessary to outpace near-peer competitors, providing data and analytical insight, and supporting an integrated budget that is aligned with strategic defense objectives.[2]

The Commission on PPBE Reform requested that the National Defense Research Institute provide an independent analysis of PPBE-like functions in selected other countries and other federal agencies. This report, part of a four-volume set, analyzes the budgeting processes of other federal agencies. Volume 1 analyzes the defense budgeting processes of China and Russia. Volume 2 analyzes the defense budgeting processes of allied countries and partners. And Volume 4, an executive summary, distills key insights from these three analytical volumes. The commission will use insights from these analyses to derive potential lessons for DoD and recommendations to Congress on PPBE reform.

This report should be of interest to those concerned with the improvement of DoD's PPBE processes. The intended audience is mostly government officials responsible for such processes. The research reported here was completed in March 2023 and underwent security review with the sponsor and the Defense Office of Prepublication and Security Review before public release.

RAND National Security Research Division

This research was sponsored by the Commission on PPBE Reform and conducted within the Acquisition and Technology Policy Program of the RAND National Security Research Divi-

---

[1] See, for example, Section 809 Panel, *Report of the Advisory Panel on Streamlining and Codifying Acquisition Regulations*, Vol. 2 of 3, June 2018, pp. 12–13; Brendan W. McGarry, *DOD Planning, Programming, Budgeting, and Execution: Overview and Selected Issues for Congress*, Congressional Research Service, R47178, July 11, 2022, p. 1; and William Greenwalt and Dan Patt, *Competing in Time: Ensuring Capability Advantage and Mission Success Through Adaptable Resource Allocation*, Hudson Institute, February 2021, pp. 9–10.

[2] Public Law 117-81, National Defense Authorization Act for Fiscal Year 2022, December 27, 2021.

sion (NSRD), which operates the National Defense Research Institute (NDRI), a federally funded research and development center sponsored by the Office of the Secretary of Defense, the Joint Staff, the Unified Combatant Commands, the Navy, the Marine Corps, the defense agencies, and the defense intelligence enterprise.

For more information on the RAND Acquisition and Technology Policy Program, see www.rand.org/nsrd/atp or contact the director (contact information is provided on the webpage).

Acknowledgments

The authors thank the members of the Commission on PPBE Reform—Robert Hale, Ellen Lord, Jonathan Burks, Susan Davis, Lisa Disbrow, Eric Fanning, Peter Levine, Jamie Morin, David Norquist, Diem Salmon, Jennifer Santos, Arun Seraphin, Raj Shah, and John Whitley—and staff for their dedication and deep expertise in shaping this work. We extend special gratitude to the commission chair, the Honorable Robert Hale; the vice chair, the Honorable Ellen Lord; executive director Lara Sayer; and director of research Elizabeth Bieri for their guidance and support throughout this analysis. We would also like to thank the subject-matter experts at the U.S. Department of Homeland Security, U.S. Department of Health and Human Services, National Aeronautics and Space Administration, and the Office of the Director of National Intelligence who provided us valuable insight on these agencies' PPBE-like processes.

From the RAND National Security Research Division, we thank Barry Pavel, vice president and director, and Mike Spirtas, associate director, along with then–acting director Christopher Mouton and associate director Yun Kang of NSRD's Acquisition and Technology Policy Program, for their counsel and tireless support. We also thank our RAND Corporation colleagues who provided input at various stages of this work, including Don Snyder, Michael Kennedy, Irv Blickstein, Brian Persons, Chad Ohlandt, Bonnie Triezenberg, Obaid Younossi, Clinton Reach, John Yurchak, Jeffrey Drezner, Brady Cillo, and Gregory Graff, as well as the team of peer reviewers who offered helpful feedback on individual case studies and on cross-case study takeaways: Lauri Rohn, Emma Westerman, Daniel Crespin, James Kallimani, Rich Girven, and Jim Powers. Finally, we would like to thank Maria Falvo and Saci Haslam for their administrative assistance during this effort. The work is much improved by virtue of their inputs, but any errors remain the responsibility of the authors alone.

Dedication

These volumes are dedicated to Irv Blickstein, whose decades of experience in the U.S. Navy's PPBE community deeply informed this work and whose intellectual leadership as a RAND colleague for more than 20 years greatly enhanced the quality of our independent analysis for DoD's most-pressing acquisition challenges. Irv's kindness, motivation, and ever-present mentoring will be sorely missed.

Summary

Issue

The U.S. Department of Defense's (DoD's) Planning, Programming, Budgeting, and Execution (PPBE) System was originally developed in the 1960s as a structured approach for planning long-term resource development, assessing program cost-effectiveness, and aligning resources to strategies. Yet changes to the strategic environment, the industrial base, and the nature of military capabilities have raised the question of whether U.S. defense budgeting processes are still well aligned with national security needs.

Congress, in its National Defense Authorization Act for Fiscal Year 2022, called for the establishment of a Commission on PPBE Reform, which took shape as a legislative commission in 2022.[1] As part of its data collection efforts, the Commission on PPBE Reform asked the National Defense Research Institute, a federally funded research and development center operated by the RAND National Security Research Division, to conduct case studies of budgeting processes across nine comparative organizations: five international defense organizations and four other U.S. federal government agencies. The two international case studies of near-peer competitors were specifically requested by Congress, while the other seven cases were selected in close partnership with the commission.

Approach

For all nine case studies, the research entailed extensive document reviews and structured discussions with subject-matter experts having experience in the budgeting processes of the selected international governments and other U.S. federal government agencies. Each case study was assigned a unique team with appropriate regional or organizational expertise. The analysis was also supplemented by experts in the U.S. PPBE process, as applicable.

Key Insights

The key insights from the case studies of selected non-DoD federal agencies—the U.S. Department of Homeland Security (DHS), the U.S. Department of Health and Human Services (HHS), the National Aeronautics and Space Administration (NASA), and the Office of the Director of National Intelligence (ODNI)—detailed in this volume are as follows:

- **Other U.S. government agencies looked to DoD's PPBE System as a model in developing their own systems, which subsequently evolved.** NASA's PPBE; ODNI's Intelligence

---

[1] Public Law 117-81, National Defense Authorization Act for Fiscal Year 2022, December 27, 2021 .

Planning, Programming, Budgeting, and Evaluation (IPPBE); DHS's PPBE; and HHS's budget process all refer to DoD's PPBE System as a model for planning and resource allocation decisionmaking. However, these agencies' budget processes have evolved differently in accordance with their missions, organizational structures, authorities, staff capacities, available resources, and many other factors. One notable and deliberate difference between ODNI's IPPBE and DoD's PPBE processes is ODNI's substitution of *evaluation* for DoD's *execution*. Despite the evolution away from DoD's PPBE framework, all four agencies still generally follow a budgeting process that is common to most U.S. federal civilian agencies. This process begins with an annual planning cycle and culminates in budget execution and performance evaluation.

- **Long-term planning is often limited relative to that done by DoD. One difference between DoD and three of the agencies considered in this report is DoD's focus on long-term planning processes.** We attribute this difference both to the inherently dynamic requirements of the DHS and HHS mission sets and to the weaker (relative to DoD) mechanisms for forging forward-looking, cross-departmental plans through a headquarters function in DHS and ODNI. Long-term planning is particularly important for agencies with missions requiring sustained development efforts rather than short-term operational programs.

- **A variety of mechanisms enable budget flexibility and agility.** Mechanisms have been designed to meet dynamic mission demands, highly variable mission needs, and emerging public health threats. Other mechanisms have given agencies more discretion (than in DoD) to redirect appropriated funds without reporting such action to Congress. HHS appears to have wide latitude in how appropriated funds are spent. Unlike DoD, NASA does not appear to receive appropriations in distinct titles. Various mechanisms allow the agencies to carry over partial funding across years to address the use-it-or-lose-it behavior associated with one-year funding, repurpose expiring unobligated balances, and reallocate funds to department-wide capital investments. In some instances, Congress further enables agility by employing broader appropriation categories than those used for DoD appropriations; in this way, agency decisionmakers have more flexibility to implement changes to previously communicated funding priorities.

- **Mechanisms for enabling agility help agencies weather continuing resolutions and other sources of budget turbulence.** Budget flexibilities can also help an agency manage under continuing resolutions. NASA's two-year expiration timeline for appropriations reportedly provides the agency with a cushion in the likely event that a regular appropriation is delayed. HHS develops requests for grant proposals ahead of anticipated continuing resolutions. The ability of DHS components to carry over into the next fiscal year (FY) up to 50 percent of prior-year balances could help the agency mitigate the effects of continuing resolutions, although that has not been the primary intention for the authority to carry over funds. Most mandatory HHS programs, such as Medicare and children's entitlement programs, are budgeted on ten-year schedules outside the annual appropriations process and, thus, are rarely subject to continuing resolutions.

- **The replacement of execution with evaluation in PPBE-like processes could be instructive for DoD.** ODNI is not alone in substituting *evaluation* for *execution* in its budgeting process. DHS has essentially done so in its PPBE-like process to better understand the results of its spending: The department now issues annual evaluation plans. This line of effort demonstrates an investment by DHS in evaluation activities. DHS's efforts in this area could help inform DoD's approach to the execution phase.
- **Implementation of PPBE-like processes at the scale of DoD's process is resource-intensive, institutionally challenging, and often infeasible for smaller agencies.** One area in which the selected non-DoD agencies cannot emulate an exemplary DoD PPBE function is DoD's Cost Assessment and Program Evaluation (CAPE) analytic function. In these four agencies, a CAPE-like function does not exist in comparable size and mission, because this function is resource-intensive to build and maintain and challenging to empower institutionally. CAPE's mission is to provide DoD with unbiased analysis on resource allocation. By comparison, the planning, programming, and budgeting for NASA are handled within one NASA organization, which may not be considered an independent organization when it scrutinizes NASA's budget submissions. ODNI attempted to emulate the analytic rigor of the CAPE function but found it difficult to do so.
- **Consolidated resource management information systems could improve visibility across the federated structures of government agencies.** DHS's consolidation of its PPBE information system has enhanced its ability to create and manage budgets. DHS officials report that the consolidated system for generating congressional budget justification documents, developing a five-year funding plan, and capturing performance management data has reduced their reliance on Microsoft Excel spreadsheet templates and data reentry, allowing DHS to automate the generation of certain reports that were previously created manually. In contrast, the lack of a consolidated budget formulation system has left HHS leadership with limited visibility into the department's operating division (OPDIV) budgets. DoD could examine the feasibility of implementing a consolidated PPBE information system and whether the benefits of doing so would outweigh the costs.

The Commission on PPBE Reform is looking for potential lessons from the PPBE-like systems of non-DoD federal agencies. Although the budgeting processes were originally modeled after DoD's PPBE System, they have adapted to the unique missions of each agency. Despite the movement away from DoD's PPBE model, the agencies still use similar planning, programming, budgeting, and execution processes. Given these similar processes, there would be no benefit from DoD adopting any of these systems wholesale. However, there is value in exploring the ways in which Congress provides each agency with flexibility so that DoD can ask for similar kinds of flexibility to support more innovation, to make funding more predictable over multiple years, and to obtain relief from various pain points in the system. These pain points include continuing resolutions, rigid appropriations categories, and appropriations for line items instead of portfolios. The commission could further explore the mechanisms for flexibility identified below, as organized by agency.

DHS funds are typically budgeted annually, but some programs receive multiyear or no-year appropriations. Congress sometimes appropriates multiyear funds to major acquisition programs to foster a stable production and contracting environment. A key example of no-year money is the Disaster Relief Fund, which is meant to give the Federal Emergency Management Agency the flexibility to respond quickly to emerging disaster relief and recovery needs. As another example, DHS officials mentioned how the border security, fencing, infrastructure, and technology appropriation gave DHS the ability to carry over significant amounts of funds related to this mission area. (DHS officials noted that funds are no longer appropriated to this account and that the use of no-year appropriations was significantly curtailed with the implementation of the common appropriations structure.) Congress also authorizes DHS components to carry one-year operations and support accounts forward into the next fiscal year and to expend up to 50 percent of the prior-year lapsed balance amounts. Beyond the base budget, DHS often receives supplemental funds for emergent requirements, the number of which varies from year to year.

HHS has access to emergency supplemental funding and several flexible-spending accounts, such as the Non-Recurring Expenses Fund, which allows HHS to reallocate expired, unobligated funds to capital investments. These flexibility mechanisms are often given multiyear or no-year funding. HHS does not use a common appropriations structure, so budget justifications focus heavily on missions and needs. This focus allows discussions between the OPDIVs and the Secretary's Budget Council's department-level leadership to concentrate on aligning program budgets and missions with the Secretary's priorities.

NASA requests and is allocated funding differently than DoD. Because NASA's funds are appropriated to mission directorates primarily at the mission, theme, and project levels, NASA has some flexibility to align project funding to meet changing priorities or real-world circumstances. Our review of NASA's FY 2023 congressional justification indicates that NASA does not request, nor is it funded with, appropriations split into categories, such as research, development, test, and evaluation (RDT&E), procurement, and operation and maintenance, in the same manner as DoD, and this was confirmed during our interviews. Therefore, NASA does not appear to encounter the same types of restrictions as DoD with respect to using specific funding for specific activities (e.g., using RDT&E only during the design and development stages of a program). Moreover, all of NASA's appropriations, except for construction, have two-year durations. NASA has obligation goals of 90–95 percent in the first year of two-year funds, which allows for some funding to be expended in the second year, typically at the start of the fiscal year. Given that continuing resolutions are a real possibility, this carryover funding can mitigate any shortfalls that might result at the start of a fiscal year—and thus act as a cushion for continuing resolutions.[2]

---

[2] Although carryover funding may mitigate some aspects of continuing resolutions, this is not its primary purpose; instead, it is designed to address the use-it-or-lose-it mentality and behavior associated with one-year funding.

ODNI funds may be reprogrammed under five conditions: (1) when funds are transferred to a high-priority intelligence activity in support of an emergent need, (2) when funds are not moved to a reserve for contingencies of the Director of National Intelligence or the Central Intelligence Agency, (3) when funds are cumulatively less than $150 million and less than 5 percent of the annual accounts available to a department or agency, (4) when the action does not terminate an acquisition program, and (5) when the congressional notification period is satisfied. Congress must be notified of above-the-threshold reprogramming actions (i.e., those that exceed $150 million or 5 percent) within 30 days, or 15 days for matters of urgent national security concern. Below-the-threshold reprogramming actions do not require congressional notification. However, ODNI does notify Congress of below-the-threshold actions that may be of congressional interest.

Contents

Figures and Tables

Figures

Tables

CHAPTER 1

Introduction

In light of a dynamic threat environment, increasingly capable adversaries, and rapid technological changes, there has been increasing concern that the U.S. Department of Defense's (DoD's) resource planning processes are too slow and inflexible to meet warfighter needs.[1] DoD's Planning, Programming, Budgeting, and Execution (PPBE) System was originally developed in the 1960s as a structured approach for planning long-term resource development, assessing program cost-effectiveness, and aligning resources to strategies. Yet changes to the strategic environment, the industrial base, and the nature of military capabilities have raised the question of whether DoD's budgeting processes are still well aligned to national security needs.

To consider the effectiveness of current resource planning processes for meeting national security needs and to explore potential policy options to strengthen those processes, Congress called for the establishment of a commission on PPBE reform in Section 1004 of the National Defense Authorization Act (NDAA) for Fiscal Year (FY) 2022.[2] The Commission on PPBE Reform took shape as a legislative commission in 2022, consisting of 14 appointed commissioners, each drawing on deep and varied professional expertise in DoD, Congress, and the private sector. In support of this work, the commission collected data, conducted analyses, and developed a broad array of inputs from external organizations, including federally funded research and development centers, to develop targeted insights of particular interest to the commission. The commission asked the RAND National Defense Research Institute to contribute to this work by conducting case studies of nine comparative organizations: five international defense organizations and four other U.S. federal government agencies. Two

---

[1] See, for example, Section 809 Panel, *Report of the Advisory Panel on Streamlining and Codifying Acquisition Regulations*, Vol. 2 of 3, June 2018, pp. 12–13; Brendan W. McGarry, *DOD Planning, Programming, Budgeting, and Execution: Overview and Selected Issues for Congress*, Congressional Research Service, R47178, July 11, 2022, p. 1; and William Greenwalt and Dan Patt, *Competing in Time: Ensuring Capability Advantage and Mission Success Through Adaptable Resource Allocation*, Hudson Institute, February 2021, pp. 9–10.

[2] Public Law 117-81, National Defense Authorization Act for Fiscal Year 2022, December 27, 2021. Section 1004 (f) of this Act is of particular relevance to our research approach:

> Compare the planning, programming, budgeting, and execution process of the Department of Defense, including the development and production of documents including the Defense Planning Guidance (described in section 113(g) of Title 10, United States Code), the Program Objective Memorandum, and the Budget Estimate Submission, with similar processes of private industry, other Federal agencies, and other countries.

of the international case studies—of near-peer competitors China and Russia—were specifically called for by Congress, and additional cases were selected in close partnership with the commission.[3]

This report is Volume 3 in a four-volume set, three of which present case studies conducted in support of the Commission on PPBE Reform. The accompanying volumes focus on selected near-peer competitors (*Planning, Programming, Budgeting, and Execution in Comparative Organizations*: Vol. 1, *Case Studies of China and Russia*) and selected U.S. partners and allies (*Planning, Programming, Budgeting, and Execution in Comparative Organizations*: Vol. 2, *Case Studies of Selected Allied and Partner Nations*).[4] Volume 4, an executive summary, distills key insights from these three analytical volumes.[5]

Evolution of DoD's PPBE System

The Planning, Programming, and Budgeting System (PPBS), the precursor to DoD's PPBE process, took shape in the first decades after World War II and was introduced into DoD in 1961 by then–Secretary of Defense Robert McNamara.[6] Drawing on new social science methods, such as program budgeting and systems analysis, the PPBS was designed to provide a structured approach to weigh the cost-effectiveness of potential defense investments. A central assertion of the PPBS's developers was that strategy and costs needed to be considered

---

[3] Pub. L. 117-81, Section 1004 (f) requires "a review of budgeting methodologies and strategies of near-peer competitors to understand if and how such competitors can address current and future threats more or less successfully than the United States."

[4] Megan McKernan, Stephanie Young, Timothy R. Heath, Dara Massicot, Mark Stalczynski, Ivana Ke, Raphael S. Cohen, John P. Godges, Heidi Peters, and Lauren Skrabala, *Planning, Programming, Budgeting, and Execution in Comparative Organizations: Vol. 1, Case Studies of China and Russia*, RAND Corporation, RR-A2195-1, 2024; Megan McKernan, Stephanie Young, Andrew Dowse, James Black, Devon Hill, Benjamin J. Sacks, Austin Wyatt, Nicolas Jouan, Yuliya Shokh, Jade Yeung, Raphael S. Cohen, John P. Godges, Heidi Peters, and Lauren Skrabala, *Planning, Programming, Budgeting, and Execution in Comparative Organizations: Vol. 2, Case Studies of Selected Allied and Partner Nations*, RAND Corporation, RR-A2195-2, 2024.

[5] See Megan McKernan, Stephanie Young, Timothy R. Heath, Dara Massicot, Andrew Dowse, Devon Hill, James Black, Ryan Consaul, Michael Simpson, Sarah W. Denton, Anthony Vassalo, Ivana Ke, Mark Stalczynski, Benjamin J. Sacks, Austin Wyatt, Jade Yeung, Nicolas Jouan, Yuliya Shokh, William Shelton, Raphael S. Cohen, John P. Godges, Heidi Peters, and Lauren Skrabala, *Planning, Programming, Budgeting, and Execution in Comparative Organizations: Vol. 4, Executive Summary*, RAND Corporation, RR-A2195-4, 2024.

[6] An oft-quoted assertion by Secretary McNamara from April 20, 1963, which is pertinent to this discussion, is that "[y]ou cannot make decisions simply by asking yourself whether something might be nice to have. You have to make a judgment on how much is enough" (as cited in the introduction of Alain C. Enthoven and K. Wayne Smith, *How Much Is Enough? Shaping the Defense Program, 1961–1969*, RAND Corporation, CB-403, 1971).

together.[7] As Charles Hitch, Secretary McNamara's first comptroller and a key intellectual leader in the development and implementation of the PPBS, noted, "There is no budget size or cost that is correct regardless of the payoff, and there is no need that should be met regardless of cost."[8]

To make decisions about prioritization and where to take risk in a resource-constrained environment, DoD needed an analytic basis for making choices. Therefore, the PPBS first introduced the program budget, an *output*-oriented articulation of the resources associated with a given military capability projected out over five years.[9] Second, the PPBS introduced an approach for assessing cost-effectiveness, termed *systems analysis*, which was institutionalized in an Office of Systems Analysis. Since 2009, this office has been known as Cost Assessment and Program Evaluation (CAPE).[10] At its inception, the PPBS was a process for explicitly linking resources to strategy and for setting up a structure for making explicit choices between options, based on transparent analysis of costs and effectiveness. Then, as today, the system introduced friction with other key stakeholders, including Congress and industry partners. Key features of the PPBS have become institutionalized in DoD's PPBE System, and questions have arisen about whether its processes and structures remain relevant and agile enough to serve their intended purposes.[11]

To set up the discussion of case studies, it will be helpful to outline the key features of the PPBE process and clarify some definitions. Figure 1.1 offers a summary view of the process.

---

[7] Or, as Bernard Brodie stated succinctly, "strategy wears a dollar sign" (Bernard Brodie, *Strategy in the Missile Age*, RAND Corporation, CB-137-1, 1959, p. 358).

[8] Charles J. Hitch and Roland N. McKean, *The Economics of Defense in the Nuclear Age*, RAND Corporation, R-346, 1960, p. 47.

[9] On the need for an output-oriented budget formulation at the appropriate level to make informed choices, Hitch and McKean (1960, p. 50) noted that the consumer "cannot judge intelligently how much he should spend on a car if he asks, 'How much should I devote to fenders, to steering activities, and to carburetion?' Nor can he improve his decisions much by lumping all living into a single program and asking, 'How much should I spend on life?'"

[10] In an essential treatise on the PPBS's founding, Enthoven (the first director of the Office of Systems Analysis) and Smith described "the basic ideas that served as the intellectual foundation for PPBS" (1971, pp. 33–47) and, thus, PPBE: (1) decisionmaking should be made on explicit criteria of the national interest, (2) needs and costs should be considered together, (3) alternatives should be explicitly considered, (4) an active analytic staff should be used, (5) a multiyear force and financial plan should project consequences into the future, and (6) open and explicit analysis should form the basis for major decisions.

[11] Greenwalt and Patt, 2021, pp. 9–10.

FIGURE 1.1

DoD's PPBE Process (as of September 2019)

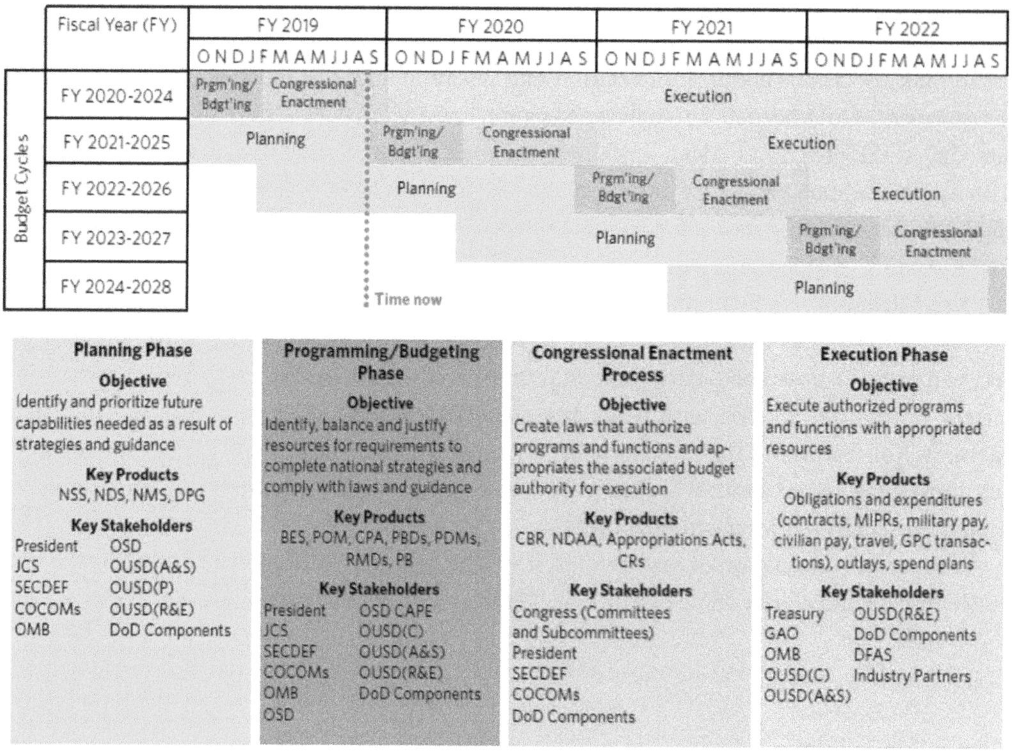

Fiscal Year (FY)	FY 2019	FY 2020	FY 2021	FY 2022
	O N D J F M A M J J A S	O N D J F M A M J J A S	O N D J F M A M J J A S	O N D J F M A M J J A S
FY 2020–2024	Prgm'ing/Bdgt'ing · Congressional Enactment	Execution		
FY 2021–2025	Planning	Prgm'ing/Bdgt'ing · Congressional Enactment	Execution	
FY 2022–2026		Planning	Prgm'ing/Bdgt'ing · Congressional Enactment	Execution
FY 2023–2027			Planning	Prgm'ing/Bdgt'ing · Congressional Enactment
FY 2024–2028	Time now			Planning

Budget Cycles

Planning Phase	Programming/Budgeting Phase	Congressional Enactment Process	Execution Phase
Objective Identify and prioritize future capabilities needed as a result of strategies and guidance	**Objective** Identify, balance and justify resources for requirements to complete national strategies and comply with laws and guidance	**Objective** Create laws that authorize programs and functions and appropriates the associated budget authority for execution	**Objective** Execute authorized programs and functions with appropriated resources
Key Products NSS, NDS, NMS, DPG	**Key Products** BES, POM, CPA, PBDs, PDMs, RMDs, PB	**Key Products** CBR, NDAA, Appropriations Acts, CRs	**Key Products** Obligations and expenditures (contracts, MIPRs, military pay, civilian pay, travel, GPC transactions), outlays, spend plans
Key Stakeholders President · OSD JCS · OUSD(A&S) SECDEF · OUSD(P) COCOMs · OUSD(R&E) OMB · DoD Components	**Key Stakeholders** President · OSD CAPE JCS · OUSD(C) SECDEF · OUSD(A&S) COCOMs · OUSD(R&E) OMB · DoD Components OSD	**Key Stakeholders** Congress (Committees and Subcommittees) President SECDEF COCOMs DoD Components	**Key Stakeholders** Treasury · OUSD(R&E) GAO · DoD Components OMB · DFAS OUSD(C) · Industry Partners OUSD(A&S)

SOURCE: Reproduced from Stephen Speciale and Wayne B. Sullivan II, "DoD Financial Management—More Money, More Problems," Defense Acquisition University, September 1, 2019, p. 6.
NOTE: BES = budget estimation submission; CBR = concurrent budget resolution; COCOM = combatant command; CPA = Chairperson's Program Assessment; CR = continuing resolution; DFAS = Defense Finance and Accounting Services; DPG = defense planning guidance; GAO = U.S. Government Accountability Office; GPC = government purchase card; JCS = Joint Chiefs of Staff; MIPR = military interdepartmental purchase request; NDS = National Defense Strategy; NMS = National Military Strategy; NSS = National Security Strategy; OMB = Office of Management and Budget; OSD = Office of the Secretary of Defense; OUSD(A&S) = Office of the Under Secretary of Defense (Acquisition and Sustainment); OUSD(C) = Office of the Under Secretary of Defense (Comptroller); OUSD(P) = Office of the Under Secretary of Defense (Policy); OUSD(R&E) = Office of the Under Secretary of Defense (Research and Engineering); PB = President's Budget; PBD = program budget decision; PDM = program decision memorandum; POM = program objective memorandum; RMD = resource management decision; SECDEF = Secretary of Defense.

Today, consideration of PPBE often broadly encapsulates internal DoD processes, other executive branch functions, and congressional rules governing appropriations. Internal to DoD, PPBE is an annual process by which the department determines how to align strategic guidance to military programs and resources. The process supports the development of DoD inputs to the President's Budget and to a budgeting program with a five-year time horizon, known as the Future Years Defense Program (FYDP).[12] DoD Directive (DoDD) 7045.14, *The Planning, Programming, Budgeting, and Execution (PPBE) Process*, states that one intent for PPBE "is to provide the DOD with the most effective mix of forces, equipment, manpower, and support attainable within fiscal constraints."[13] PPBE consists of four distinct processes, each with its own outputs and stakeholders. Select objectives of each phase include the following:

- **Planning:** "Integrate assessments of potential military threats facing the country, overall national strategy and defense policy, ongoing defense plans and programs, and projected financial resources into an overall statement of policy."[14]
- **Programming:** "[A]nalyze the anticipated effects of present-day decisions on the future force" and detail the specific forces and programs proposed over the FYDP period to meet the military requirements identified in the plans and within the financial limits.[15]
- **Budgeting:** "[E]nsure appropriate funding and fiscal controls, phasing of the efforts over the funding period, and feasibility of execution within the budget year"; restructure budget categories for submission to Congress according to the appropriation accounts; and prepare justification material for submission to Congress.[16]
- **Execution:** "[D]etermine how well programs and financing have met joint warfighting needs."[17]

Several features of congressional appropriations processes are particularly important to note. First, since FY 1960, Congress has provided budget authority to DoD through specific appropriations titles (sometimes termed *colors of money*), the largest of which are operation and maintenance (O&M); military personnel; research, development, test, and evaluation (RDT&E); and procurement.[18] These appropriations titles are further broken down into *appropriation accounts*, such as Military Personnel, Army or Shipbuilding and Conversion, Navy (SCN). Second, the budget authority provided in one of these accounts is generally available for obligation only within a specified period. In the DoD budget, the period of availability for military personnel and O&M accounts is one year; for RDT&E accounts, two years; and for most procurement accounts, three years (although for SCN, it can be five or six years,

---

[12] Brendan W. McGarry, *Defense Primer: Planning, Programming, Budgeting and Execution (PPBE) Process*, Congressional Research Service, IF10429, January 27, 2020, p. 1.

[13] DoDD 7045.14, *The Planning, Programming, Budgeting, and Execution (PPBE) Process*, U.S. Department of Defense, August 29, 2017, p. 2.

[14] Congressional Research Service, *A Defense Budget Primer*, RL30002, December 9, 1998, p. 27.

[15] Congressional Research Service, 1998, p. 27; McGarry, 2020, p. 2.

[16] McGarry, 2020, p. 2; Congressional Research Service, 1998, p. 28.

[17] DoDD 7045.14, 2017, p. 11.

[18] Congressional Research Service, 1998, pp. 15–17.

in certain circumstances). This specification means that budget authority must be obligated within those periods or, with only a few exceptions, it is lost.[19] There has been recent interest in exploring how these features of the appropriations process affect transparency and oversight, institutional incentives, and the exercise of flexibility, should resource needs change.[20]

Importantly, PPBE touches almost everything DoD does and, thus, forms a critical touchpoint for engagement with stakeholders across DoD (e.g., OSD, military departments, Joint Staff, COCOMs), in the executive branch (through OMB), in Congress, and among industry partners.

Research Approach and Methods

In close partnership with the commission, we selected nine case studies to explore decision-making in organizations facing challenges similar to those experienced in DoD: exercising agility in the face of changing needs and enabling innovation. Two near-peer case studies were specifically called for in the legislation, in part to allow the commission to explore the competitiveness implications of strategic adversaries' approaches to resource planning.

For all nine case studies, we conducted extensive document reviews and structured discussions with subject-matter experts having experience in the budgeting processes of the international governments and other U.S. federal government agencies. For case studies of two allied and partner countries, the team leveraged researchers in RAND Europe (located in Cambridge, United Kingdom) and RAND Australia (located in Canberra, Australia) with direct experience in partner defense organizations. Given the diversity in subject-matter expertise required across the case studies, each one was assigned a unique team with appropriate regional or organizational expertise. For the near-peer competitor cases, the assigned experts had the language skills and methodological training to facilitate working with primary sources in Chinese or Russian. The analysis was also supplemented by experts in PPBE as applicable.

Case study research drew primarily on government documentation outlining processes and policies, planning guidance, budget documentation, and published academic and policy research. Although participants in structured discussions varied in accordance with the decisionmaking structures across case studies, they generally included chief financial officers, representatives from organizations responsible for making programmatic choices, and budget officials. For obvious reasons, the China and Russia case studies faced unique challenges in data collection and in identifying and accessing interview targets with direct knowledge of PPBE-like processes.

---

[19] Congressional Research Service, 1998, pp. 49–50. Regarding RDT&E, see U.S. Code, Title 10, Section 3131, Availability of Appropriations.

[20] McGarry, 2022.

To facilitate consistency, completeness in addressing the commission's highest-priority areas of interest, and cross-case comparisons, the team developed a common case study template. This template took specific questions from the commission as several inputs, aligned key questions to PPBE processes and oversight mechanisms, evaluated perceived strengths and challenges of each organization's processes and their applicability to DoD processes, and concluded with lessons learned from each case. To enable the development of a more consistent evidentiary base across cases, the team also developed a standard interview protocol to guide the structured discussions.

Areas of Focus

Given the complexity of PPBE and its many connections to other processes and stakeholders, along with other inputs and ongoing analysis by the commission, we needed to scope this work in accordance with three of the commission's top priorities.

First, although we sought insights across PPBE phases in each case study, in accordance with the commission's guidance, we placed a particular emphasis on an organization's budgeting and execution mechanisms, such as the existence of appropriations titles (i.e., colors of money), and on any mechanisms for exercising flexibility, such as reprogramming thresholds. However, it is important to note that this level of detailed information was not uniformly available. The opacity of internal processes in China and Russia made the budget mechanisms much more difficult to discern in those cases in particular.

Second, while the overall investment portfolios varied in accordance with varying mission needs, the case studies were particularly focused on investments related to RDT&E and procurement rather than O&M or sustainment activities.

Third, the case studies of other U.S. federal government agencies did not focus primarily on the roles played by external stakeholders, such as OMB, Congress, and industry partners. Such stakeholders were discussed when relevant insights emerged from other sources, but interviews and data collection were focused within the bounds of a given organization rather than across a broader network of key stakeholders.

Research Limitations and Caveats

This research required detailed analysis of the nuances of internal resource planning processes across nine extraordinarily diverse organizations and on a tight timeline required by the commission's challenging mandate. This breadth of scope was intended to provide the commission with diverse insights into how other organizations address similar challenges but also limited the depth the team could pursue for any one case. These constraints warrant additional discussion of research limitations and caveats of two types.

First, each case study, to a varying degree, confronted limitations in data availability. The teams gathered documentation from publicly available sources and doggedly pursued additional documentation from targeted interviews and other experts with direct experience, but even for the cases from allied countries and U.S. federal agencies, including DoD, there was a limit to what could be established in formal documentation. Some important features of how systems work in practice are not captured in formal documentation, and such features had to be teased out and triangulated from interviews to the extent that appropriate officials were available to engage with the team. The general opacity and lack of institutional connections to decisionmakers in China and Russia introduced unique challenges for data collection. Russia was further obscured by the war in Ukraine during the research period, which made access by U.S.-based researchers to reliable government data on current plans and resource allocation impossible.

Second, the case study teams confronted important inconsistencies across cases, which made cross-case comparability very challenging to establish. For example, international cases each involved unique political cultures, governance structures, strategic concerns, and military commitments—all of which we characterize to the extent that it is essential context for understanding how and why resource allocation decisions are made. The context-dependent nature of the international cases made even defining the "defense budget" difficult, given countries' various definitions and inclusions. With respect to the near-peer case studies, inconsistencies were especially pronounced regarding the purchasing power within those two countries. To address some of these inconsistencies, we referenced the widely cited Stockholm International Peace Research Institute (SIPRI) Military Expenditure Database.[21] With respect to the other U.S. federal agencies, each agency had its own unique mission, organizational culture, resource level, and process of congressional oversight—all of which were critical for understanding how and why resource allocation decisions were made. This diversity strained our efforts to draw cross-case comparisons or to develop internally consistent normative judgments of best practices. For this reason, each case study analysis and articulation of strengths and challenges should be understood relative to each organization's *own* unique resource allocation needs and missions.

Selected Non-DoD Federal Agencies Focus

The 2022 NDS describes a security environment of complex strategic challenges associated with such dynamics as emerging technology, transboundary threats, and competitors posing "new threats to the U.S. homeland and to strategic stability."[22] To meet this challenge, the NDS calls on DoD to undertake three activities: integrated deterrence, campaigning, and "build[ing] enduring advantage." The last category is defined as "undertaking reforms to accelerate force development, getting the technology we need more quickly, and making investments in the extraordinary people of the Department, who remain our most valuable resource."[23] This imperative has prompted reflection on the extent to which internal DoD

---

[21] SIPRI, "SIPRI Military Expenditure Database," homepage, undated.

[22] DoD, *2022 National Defense Strategy of the United States of America*, 2022, p. 4.

[23] DoD, 2022, p. iv.

processes, including PPBE, are up to the challenge of enabling rapid and responsive capability development to address the emerging threats.

Notably, the idea of dialogue between DoD and non-DoD agencies for lessons in resource planning areas is not new; indeed, in 1965, President Lyndon B. Johnson decided to introduce the still-new DoD PPBS across the federal government. Two of the cases considered in this volume, U.S. Department of Health and Human Services (HHS) and National Aeronautics and Space Administration (NASA), were included in this 1965 directive before the experiment fizzled out in 1970. The other two cases considered in this report—Office of the Director of National Intelligence (ODNI) and U.S. Department of Homeland Security (DHS)—also have PPBE-like functions that notably resonate with PPBE's origins in DoD. Although Johnson's mandate was relatively short-lived, all four of the case studies considered in this report looked to DoD's PPBE process in the development of their own processes. As these relatively new cabinet departments set up organizations with strong existing processes and organizational cultures, they were grappling with somewhat similar challenges as the then–relatively new DoD on issues related to strategic planning, enterprise decisionmaking, and institutional control. In this context, the development of PPBE-like processes in these agencies was an important effort for institutional development. Figure 1.2 provides a summary view of the

FIGURE 1.2

Discretionary Budget Authority by U.S. Government Agency, 2021

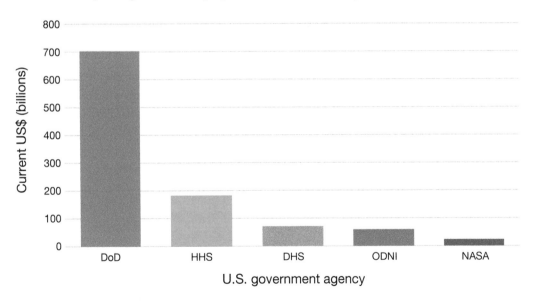

SOURCES: Features information from OMB, "Historical Tables: Table 5.4—Discretionary Budget Authority by Agency: 1976–2028," White House, undated; ODNI, "U.S. Intelligence Community Budget," webpage, undated-f.

NOTES: For ODNI, we show the total budget appropriation for the National Intelligence Program (NIP) and not the discretionary budget authority because of a lack of available data for comparison in the OMB tables. HHS's total annual budget authority was nearly $1.7 trillion in 2021, which was more than double DoD's total 2021 budget authority of roughly $720 billion. Mandatory HHS funding (primarily for Medicare and Medicaid) constitutes about 90 percent of the total HHS budget; however, only its discretionary budget authority is presented here.

budgets of the four selected federal agencies, compared with DoD, in 2021—of which, DoD had the largest discretionary budget authority of the five agencies.

Beyond these historical observations, the four other U.S. government agencies selected for analysis were identified as agencies that, by virtue of their missions, grappled with some issues similar to those that DoD faced (and continues to face) regarding how to enable innovation, make high-tech investments, and transition technology or remain flexible in light of dynamic mission needs. Although each agency is different from DoD in important ways, their unique stories also provide some notable insights for the commission. We provide introductory overviews for each agency in the following sections, as drawn from the respective case studies in Chapters 2 through 5.

U.S. Department of Homeland Security

DHS, established in 2002, protects the U.S. homeland through a broad array of missions, including border, transportation, and maritime security; cyber and infrastructure security; management of the immigration system and immigration enforcement; disaster response; protection of the nation's leaders; and countering weapons of mass destruction.[24] Because of the unpredictable nature of these missions, DHS must be able to react to changing conditions.

Prior to the formation of DHS, 22 federal departments and agencies performed homeland security–type missions.[25] As shown in Figure 1.3, DHS contains eight operational components (blue) and seven support offices and directorates (green). Recent major organizational changes created the Cybersecurity and Infrastructure Security Agency (CISA) as an operational component and the Countering Weapons of Mass Destruction (CWMD) Office as a support component.

DHS is the third-largest federal cabinet-level agency, with more than 250,000 employees and nearly $100 billion in the FY 2023 President's Budget request.[26] The annual DHS budget includes discretionary funding for emerging disasters (the Disaster Relief Fund), which has historically accounted for anywhere between 20 and 60 percent of the total budget; discretionary fees; and mandatory appropriations, such as personnel, compensation, and benefits. DHS's budget also includes billions of dollars for major acquisition programs in support of its critical missions. These acquisition programs include a variety of systems to secure the border, secure cyberspace, screen travelers, and conduct other activities.[27] In addition to the base budget, DHS often receives supplemental funds for emergent requirements.

---

[24] For more information, see DHS, "Department of Homeland Security's Strategic Plan for Fiscal Years 2020–2024," webpage, undated-a.

[25] DHS, "History," webpage, last updated April 26, 2022a.

[26] DHS, *U.S. Department of Homeland Security Agency Financial Report, FY 2022*, November 15, 2022b, p. 1.

[27] See GAO, *DHS Annual Assessment: Most Acquisition Programs Are Meeting Goals Even with Some Management Issues and COVID-19 Delays*, GAO-22-104684, March 8, 2022. Because of the significant life-cycle

FIGURE 1.3
DHS Organizations

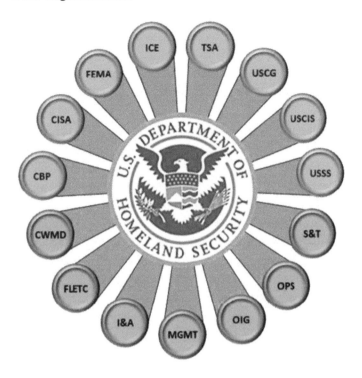

SOURCE: DHS, 2022b.
NOTES: Organizations in blue are DHS components, and organizations in green are support offices and directorates.
CBP = U.S. Customs and Border Protection; CISA = Cybersecurity and Infrastructure Security Agency; FEMA = Federal Emergency Management Agency; ICE = U.S. Immigration and Customs Enforcement; TSA = Transportation Security Administration; USCG = U.S. Coast Guard; USCIS = U.S. Citizenship and Immigration Services; USSS = U.S. Secret Service; CWMD = Countering Weapons of Mass Destruction Office; FLETC = Federal Law Enforcement Training Center; I&A = Office of Intelligence and Analysis; MGMT = Management Directorate; OIG = Office of Inspector General; OPS = Office of Operations Coordination (now known as the Office of Homeland Security Situational Awareness, or OSA); S&T = Science and Technology Directorate.

costs associated with major acquisition programs, DHS has implemented policies to oversee the management of these programs to help ensure that they meet cost, schedule, and performance requirements, and DHS maintains mechanisms to track these areas, such as Acquisition Program Health Assessments. Prior to

Prior to 2002, each of the original 22 national security–related organizations had its own budget. When DHS was formed, Congress provided it with budget flexibilities, such as *no-year* accounts, to give it some latitude as it was stood up to allow the organization to mature.[28] However, DHS faced significant challenges with its financial management because it inherited an array of legacy component systems and processes.[29] Additionally, DHS leadership under former secretaries Tom Ridge, Michael Chertoff, and Janet Napolitano recognized a need to increase cohesion within the department and implemented an initiative known as "One DHS."[30]

Under former Secretary Jeh Johnson (2013–2017), DHS sought to institutionalize decision-making bodies and add requirements processes in a "unity-of-effort initiative," thereby seeking to develop the department's PPBE system.[31] However, DHS's federated model, in which the components remain responsible for their separate missions and receive direct appropriations, makes it more difficult to promote departmental priorities, unlike in DoD where the Secretary of Defense has more control over the military services and agencies. DHS lacks a Goldwater-Nichols initiative to compel jointness, and DHS headquarters lacks the resourcing of OSD, which limits its ability to conduct coordination and management.

Because of a rapidly evolving threat landscape and a politically charged environment, DHS budgeting focuses largely on near-term challenges, and the department's strategic planning capabilities remain immature.[32] However, the department has obtained a clean audit opinion for ten consecutive years,[33] and it recently implemented a consolidated, department-

---

becoming acquisition programs, they must undergo a DHS process to review and validate requirements to identify synergies across the department and help ensure that acquisition programs fill gaps in capabilities to accomplish DHS's mission. DHS tracks the extent to which capability gaps have been filled.

[28] DHS officials, interviews with the authors, September–November 2022.

[29] See, for example, McCoy Williams, "Department of Homeland Security: Financial Management Challenges," testimony before the Subcommittee on Financial Management, the Budget, and International Security, Committee on Governmental Affairs, U.S. Senate, GAO-04-945T, U.S. General Accounting Office, July 8, 2004.

[30] See William L. Painter, Michael E. DeVine, Bart Elias, Kristin Finklea, John Frittelli, Jill C. Gallagher, Frank Gottron, Diane P. Horn, Chris Jaikaran, Lennard G. Kruger, et al., *Selected Homeland Security Issues in the 116th Congress*, Congressional Research Service, R45701, November 26, 2019.

[31] For more information, see Jeh Johnson, Secretary of Homeland Security, "Strengthening Departmental Unity of Effort," memorandum for DHS leadership, April 22, 2014.

[32] DHS official, interview with the authors, September 2022. Also see GAO, *Homeland Security: Clearer Roles and Responsibilities for the Office of Strategy, Policy, and Plans and Workforce Planning Would Enhance Its Effectiveness*, GAO-18-590, September 2018; and GAO, *Quadrennial Homeland Security Review: Improved Risk Analysis and Stakeholder Consultations Could Enhance Future Reviews*, GAO-16-371, April 2016a.

[33] Obtaining a *clean opinion* means that DHS has presented its financial statements fairly and in accordance with generally accepted accounting principles. See GAO, *Department of Homeland Security: Progress Made Strengthening Management Functions, but Work Remains*, GAO-21-105418, September 30, 2021; and DHS

wide PPBE system to harmonize information into a single system of record.[34] DoD could learn from DHS's simplified strategic planning approach, its evaluation efforts, and its consolidated PPBE data system.

U.S. Department of Health and Human Services

HHS, established with its current organizational structure and mission in 1980, provides health and human services and pursues scientific advances in the areas of medicine, public health, and social services. The department has five primary strategic goals: protecting and strengthening equitable access to high-quality and affordable health care; safeguarding and improving national and global health outcomes; strengthening social well-being, equity, and economic resilience; restoring trust and accelerating progress in science and research; and advancing strategic management to build trust, transparency, and accountability.[35]

HHS traces its roots as a federal agency to the creation of the Federal Security Agency in 1939 and the subsequent elevation of the U.S. Department of Health, Education, and Welfare (HEW) to cabinet-level status in 1953.[36] Since then, the department has undergone several major reorganizations, including the U.S. Department of Education's elevation to a stand-alone, cabinet-level department in 1979; HEW's renaming to the U.S. Department of Health and Human Services in 1980; and the Social Security Administration's spin-off as an independent agency in 1994. As shown in Table 1.1, HHS has 12 operating divisions (OPDIVs) and 14 Office of the Secretary staff divisions. The OPDIVs are responsible for administering HHS's Public Health Service missions,[37] such as protecting the U.S. public from pathogens and other public health threats, conducting health and biomedical research, and promoting the economic and social well-being of U.S. communities. Staff divisions provide support to the Office of the Secretary of Health and Human Services in administering all HHS programs and activities.

A decade after its 1953 elevation to a cabinet-level department, HEW adopted an early version of PPBE as part of the Johnson administration's effort in 1965 to establish processes for program budgeting in all U.S. federal civilian agencies. HEW modeled its program budgeting framework and organizational structure after DoD's PPBE model, which included creating an office (ASPE) to oversee evaluation and budget trade-offs and linking strategic planning efforts to five-year budget plans.[38] Although some vestiges of this framework—such as its rig-

---

OIG, *Independent Auditors' Report on the Department of Homeland Security's Consolidated Financial Statements for FYs 2022 and 2021 and Internal Control over Financial Reporting*, OIG-23-02, November 15, 2022.

[34] DHS official, interview with the authors, October 2022.

[35] For more information, see HHS, "Strategic Plan FY 2022–2026," webpage, last reviewed March 28, 2022a.

[36] HHS, "HHS Historical Highlights," webpage, last reviewed March 14, 2023a.

[37] Nine of the 12 OPDIVS formally constitute the U.S. Public Health Service agencies.

[38] Alice M. Rivlin, "The Planning, Programming, and Budgeting System in the Department of Health, Education, and Welfare: Some Lessons from Experience," in U.S. House of Representatives, Joint Economic

TABLE 1.1

HHS Organizations

Operating Divisions	Staff Divisions
• Administration for Children and Families • Administration for Community Living • Agency for Healthcare Research and Quality • Administration for Strategic Preparedness and Response • Agency for Toxic Substances and Disease Registry • Centers for Disease Control and Prevention (CDC) • Centers for Medicare and Medicaid Services • Food and Drug Administration (FDA) • Health Resources and Services Administration • Indian Health Service • National Institutes of Health (NIH) • Substance Abuse and Mental Health Services Administration (SAMHSA)	• Office of the Assistant Secretary for Administration (ASA) • Office of the Assistant Secretary for Financial Resources (ASFR) • Office of the Assistant Secretary for Health • Office of the Assistant Secretary for Legislation • Office of the Assistant Secretary for Planning and Evaluation (ASPE) • Office of the Assistant Secretary for Public Affairs • Office for Civil Rights • Departmental Appeals Board • Office of the General Counsel • Office of Global Affairs • Office of Inspector General (HHS OIG) • Office of Medicare Hearings and Appeals • Office of the National Coordinator for Health Information Technology • Chief Information Officer

SOURCE: Features information from HHS, "HHS Organizational Charts: Office of Secretary and Divisions," webpage, August 17, 2023b.

orous program evaluation capabilities—remain features of the contemporary HHS budgeting system, the department gradually dismantled much of its earlier PPBS during the 1970s in response to the perception that PPBS did not fit with HEW's missions, organizational structure, or program needs.[39] Despite its shared legacy with DoD's PPBE, HHS's budgeting system has therefore diverged significantly from DoD's since 1980.

Because HHS programs focus on delivering health care services and grants, outside its mandatory funding, the department operates primarily on one-year discretionary funding and restricts budget planning to the annual budget cycle.[40] Consequently, HHS does not engage in robust long-term budget planning, nor does it have well-established links between strategic planning and budgeting.[41] However, the department benefits from considerable budgetary flexibility, which Congress provides so that HHS can respond to unpredictable

---

Committee, Subcommittee on Economy in Government, *The Analysis and Evaluation of Public Expenditures: The PPB System, A Compendium of Papers Submitted to the Subcommittee on Economy in Government of the Joint Economic Committee*, Vol. 3, Part V, Section C, U.S. Government Printing Office, 1969, p. 911.

[39] See, for example, Robert L. Harlow, "On the Decline and Possible Fall of PPBS," *Public Finance Quarterly*, Vol. 1, No. 2, April 1973, p. 90; Stephen F. Jablonsky and Mark W. Dirsmith, "The Pattern of PPB Rejection: Something About Organizations, Something About PPB," *Accounting, Organizations and Society*, Vol. 3, Nos. 3–4, 1978, p. 216; Rivlin, 1969, p. 922; and U.S. General Accounting Office, *Management of HHS: Using the Office of the Secretary to Enhance Departmental Effectiveness*, GAO-HRD-90-54, February 9, 1990, p. 22.

[40] HHS officials, interviews with the authors, October 2022–January 2023.

[41] HHS officials, interviews with the authors, October 2022–January 2023.

mission needs. The department also benefits from consistent policies and processes, effective mechanisms for adjudicating budget priorities and trade-offs, relative effectiveness in managing continuing resolutions, and strong transparency and oversight mechanisms. DoD could learn from HHS's collaborative top-down and bottom-up budgeting process, its budget flexibility, and its centralized oversight mechanisms.

National Aeronautics and Space Administration

NASA was established in 1958 and is responsible for U.S. aeronautical and space activities.[42] As a civilian agency, NASA is devoted to peaceful activities that benefit humankind. It describes its mission as follows: "NASA explores the unknown in air and space, innovates for the benefit of humanity, and inspires the world through discovery."[43] The National Aeronautics and Space Act of 1958 outlined eight activities that support this mission,[44] which we enumerate in Chapter 4. According to NASA, these activities are supported by its core values of safety, integrity, teamwork, excellence, and inclusion.[45]

NASA has more than 18,000 civil servants from a multitude of disciplines with various backgrounds who work with many more contractors, academics, international colleagues, and commercial partners to accomplish its mission. Overall, NASA's workforce comprises more than 312,000 professionals at 20 centers and facilities across the United States.[46] Figure 1.4 shows NASA's organizational and reporting structure.

NASA's core values and attributes appear to influence its approaches to PPBE—especially its team-oriented processes. NASA's culture is highly collegial, which promotes engagement up and down the organization. The entire budgeting process—from development to execution and performance management—is overseen by the Office of the Chief Financial Officer (OCFO) through representatives at NASA headquarters, the Mission Support Directorate, five mission directorates, and ten geographically dispersed centers, including the Jet Propulsion Laboratory, NASA's federally funded research and development center.[47] NASA's annual budget generally amounts to less than 0.5 percent of the entire U.S. federal budget, and it received $23.27 billion in funding in FY 2021. By comparison, NASA generated economic output of more than $71.2 billion in FY 2021, and this output resulted in approximately $7.7 billion in federal, state, and local tax revenues that year.[48]

---

[42] Public Law 85-568, National Aeronautics and Space Act of 1958, July 29, 1958.

[43] NASA, "Missions," webpage, last updated April 15, 2022e.

[44] Pub. L. 85-568, 1958.

[45] NASA, 2022e.

[46] NASA, "About NASA," webpage, last updated January 27, 2023a.

[47] NASA's PPBE process is outlined in a single publicly available document: NASA Procedural Requirement (NPR) 9420.1A, *Budget Formulation*, Office of the Chief Financial Officer, National Aeronautics and Space Administration, incorporating change 1, September 15, 2021.

[48] NASA, *Economic Impact Report*, October 2022f.

FIGURE 1.4
NASA's Organizational Structure

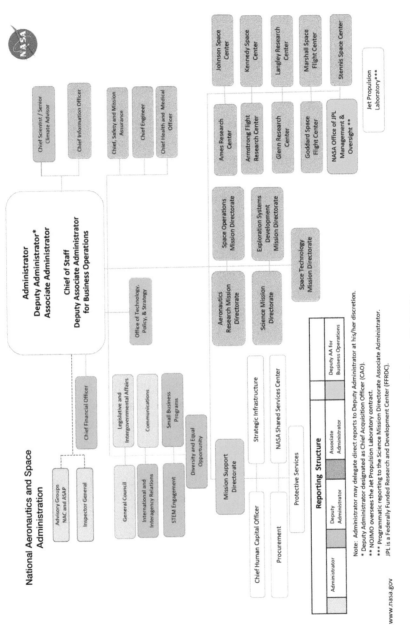

SOURCE: Reproduced from NASA, organizational chart, February 25, 2022d.

NOTES: AA = Associate Administrator; ASAP = Aerospace Safety Advisory Panel; JPL = Jet Propulsion Laboratory; NAC = NASA Advisory Council; NOJMO = Office of JPL Management and Oversight.

NASA's budgeting process has several perceived strengths and challenges for meeting mission needs. As a strength, despite NASA's comparatively smaller annual budget, several features of its budgeting mechanisms enable a degree of flexibility and agility. For example, all appropriations are for two years, except for construction, which is funded on a six-year basis. Appropriation categories also appear more fungible, with fewer restrictions on how appropriated funds are spent. Funding is also budgeted primarily at the program level or above, which we interpreted as providing NASA with broader appropriation categories within which to shift funding in accordance with changing priorities. Relative to DoD, NASA also seems to have more-streamlined processes that are closely associated with PPBE, such as its requirements development processes. However, all PPBE functions are also focused in the OCFO, which is a narrower basis for stakeholder input, and there is limited institutional capability for analytic functions, similar to the role played by CAPE in DoD's PPBE.

Office of the Director of National Intelligence

After the terrorist attacks on September 11, 2001, intelligence integration became a principal concern for the federal government. The Intelligence Reform and Terrorism Prevention Act of 2004 (IRTPA) was designed to promote the operational integration of the U.S. intelligence community (IC) through the creation of ODNI as the IC coordinator and to improve information-sharing across the IC.[49] Prior to IRTPA, the director of the Central Intelligence Agency (CIA) had been dual-hatted, also acting as the nominal head of the IC, with the assistance of a small community of management staff within CIA headquarters.[50]

ODNI began operations in 2005. Its early years were hampered by instability in staffing the roles of Director of National Intelligence (DNI) and Principal Deputy Director of National Intelligence (PDDNI). From 2005 to 2010, the DNI position was filled by John Negroponte, Mike McConnell, Dennis Blair, and James Clapper. Clapper's longer tenure than his predecessors, from 2010 to 2017, provided ODNI with an opportunity to establish itself within the IC. Moreover, Clapper provided an overarching vision, rationale, and value proposition for the office's relationship with the IC and the concept of vertical and horizontal integration of the IC and its activities.[51]

Additionally, Clapper and his PDDNI, Stephanie O'Sullivan, built on the work that ODNI staff had begun under Dennis Blair to piece together the IC budget process, which would be

[49] Public Law 108-458, Intelligence Reform and Terrorism Prevention Act of 2004, December 17, 2004; Jon Rosenwasser, "Intelligence Integration: A Congressional Oversight Perspective," *Studies in Intelligence* (extracts), Vol. 65, No. 3, September 2021. This view of ODNI's role was also shared by an ODNI subject-matter expert in an interview with the authors in February 2023.

[50] ODNI subject-matter experts, interviews with the authors, November–December 2022; Director of Central Intelligence Directive 3/3, *Community Management Staff*, Central Intelligence Agency, June 12, 1995.

[51] ODNI subject-matter experts, interviews with the authors, November 2022–February 2023; Jim Clapper and Trey Brown, "A DNI's Overview: Reflection on Integration in the Intelligence Community," *Studies in Intelligence* (extracts), Vol. 65, No. 3, September 2021.

codified in 2011 as Intelligence Community Directive (ICD) 116.[52] Clapper and O'Sullivan then used ICD 116 as a key component for integrating the budget processes across the IC.

ODNI's mission is to lead intelligence integration across the intelligence enterprise and forge an IC that gives policymakers and warfighters a decisionmaking advantage by consistently and routinely delivering the most insightful intelligence possible.[53] The IC is a coalition of 18 agencies and organizations—specifically, the IC consists of two independent organizations (ODNI and CIA); nine DoD elements (Defense Intelligence Agency, National Security Agency, National Geospatial-Intelligence Agency, National Reconnaissance Office, U.S. Air Force, U.S. Army, U.S. Marine Corps, U.S. Navy, and U.S. Space Force); and seven elements of other departments and agencies (U.S. Coast Guard Intelligence; U.S. Department of Energy, Office of Intelligence and Counterintelligence; DHS, Intelligence and Analysis; U.S. Department of Justice, Drug Enforcement Agency, Office of National Security Intelligence; U.S. Department of Justice, Federal Bureau of Investigation (FBI); U.S. Department of State, Bureau of Intelligence and Research; and U.S. Department of Treasury, Office of Intelligence and Analysis).[54] Figure 1.5 depicts all 18 departments and agencies that make up the IC (with ODNI in the center).

Since 2007, the number of military and government civilian employees at ODNI has remained relatively consistent. Of the office's approximately 1,700 personnel, 40 percent are on rotation from the other 17 IC elements.[55] The National Counterterrorism Center, the National Counterproliferation and Biosecurity Center, the National Counterintelligence and Security Center, and the Foreign Malign Influence Center are all part of ODNI.[56] Unlike other headquarters elements, such as those within DoD, ODNI employs fewer contractors than government and military staff.

IRTPA gives the DNI primary responsibility for the following activities, several of which broadly relate to resource planning processes:[57]

- serving as the head of the IC
- acting as principal adviser for intelligence matters related to national security

---

[52] ICD 116, *Intelligence Planning, Programming, Budgeting, and Evaluation System*, Office of the Director of National Intelligence, September 14, 2011.

[53] ODNI, "Mission, Vision & Values," webpage, undated-d.

[54] ODNI, "What We Do," webpage, undated-g.

[55] ODNI subject-matter experts, interviews with the authors, August 2022–February 2023; ODNI, "ODNI Factsheet," February 24, 2017.

[56] ODNI subject-matter experts, interviews with the authors, November 2022–February 2023; Mark M. Lowenthal, *Intelligence: From Secrets to Policy*, 8th ed., CQ Press, 2019, p. 42; ODNI, organizational chart, October 20, 2022.

[57] Pub. L. 108-458, 2004.

FIGURE 1.5

U.S. Intelligence Community Elements

SOURCE: Reproduced from ODNI, undated-g.

- directing the execution of the NIP, developing an annual consolidated budget for the NIP, and managing the NIP's appropriated funds
- participating in the development of the annual budget for the Military Intelligence Program (MIP).

The DNI is authorized to perform the following budgetary functions:[58]

- provide guidance to IC components regarding the development of the NIP based on intelligence priorities established by the President
- develop and determine an annual consolidated NIP budget
- present a finalized budget for presidential approval
- transfer and reprogram funds within the NIP with the approval of the director of OMB and in consultation with the affected agencies.

The NIP and the MIP are the two major components of the U.S. intelligence budget. The NIP "includes all programs, projects, and activities of the intelligence community as well as any other intelligence community programs designated jointly by the DNI and the head of department or agency, or the DNI and the President."[59] In contrast, the MIP is "devoted to intelligence activity conducted by the military departments and agencies in the Department of Defense that support tactical U.S. military operations."[60] The MIP is managed through the standard DoD PPBE process by the Under Secretary of Defense for Intelligence and Security.[61]

ODNI manages the NIP through the Intelligence Planning, Programming, Budgeting, and Evaluation (IPPBE) System (also the title of ICD 116). The IPPBE process was specifically modeled on DoD's PPBE process, with some modifications to ensure that the IPPBE suited ODNI's mission. This PPBE adaptation was done both by necessity, to account for the presence of DoD elements in the IC, and by design, because the architects of the IPPBE process were well versed in PPBE and viewed it as the best model for ODNI's complex organizational structure.[62] One notable and deliberate difference between the IPPBE and PPBE processes is ODNI's substitution of *evaluation* for DoD's *execution*.[63] The authors of ICD 116 put in place a comprehensive evaluation system to help shape the future allocation of resources within the IPPBE process.[64]

The mechanisms of the ODNI budgeting process have evolved with changing leadership priorities, resourcing constraints, and staff reorganizations, but there are several perceived strengths of ODNI's IPPBE process to note. It evolved out of the DoD PPBE framework but with an increased emphasis on various evaluation mechanisms that, together, support con-

---

[58] Pub. L. 108-458, 2004.

[59] ODNI, undated-f.

[60] ODNI, undated-f.

[61] DoDD 5205.12, *Military Intelligence Program*, change 2, October 1, 2020.

[62] ODNI subject-matter experts, interviews with the authors, August–September 2022.

[63] ODNI subject-matter experts, interviews with the authors, August–September 2022.

[64] ODNI subject-matter experts, interviews with the authors, August–September 2022. Also see ICD 116, 2011. The accountability and evaluation tools in the IPPBE process have changed over time, but they have included strategic evaluation reports (SERs), major issue studies (MISs), consolidated intelligence guidance (CIG) compliance reports, and strategic program briefings.

tinuous evaluation and an opportunity to incorporate feedback from multiple parts of the staff. IPPBE includes mechanisms for establishing strategic priorities and synchronizing staff inputs and for integrating the priorities of the IC overall. Yet within this structure, the process has proved flexible and adaptable for managing a complex network of 18 disparate intelligence organizations and has endured through multiple ODNI staff reorganizations.

Yet IPPBE also faces some perceived challenges in execution, in part due to the exceptional organizational complexity of the IC, the IC budget, and challenges related to synchronization and alignment, as well as limited staff capacity within ODNI. For example, while the "NIP-MIP Rules of the Road" for ODNI's budgeting process for the NIP are established,[65] and while DoD's management of the MIP is clearly documented under the PPBE process, the relationship between ODNI and non-DoD IC agencies can be more complex. In addition, execution of the IPPBE process has evolved in a relatively short time, which can lead to a sense that the rules are constantly in flux. The ODNI staff overseeing the IPPBE process is relatively small for the scope, scale, and complexity of the mission. This limited capacity extends to ODNI's capability and capacity to provide timely and sophisticated analytic support to decisionmakers, analogous to the role that CAPE plays in DoD.

Structure of This Report

Chapter 2 provides a detailed case study on DHS's resource planning, followed by Chapter 3 on HHS's resource planning. Chapter 4 is a case study on NASA's resource planning, and Chapter 5 is the final case study, which is on ODNI's resource planning. Chapter 6 reviews key insights across the four case studies.

---

[65] ODNI, "NIP-MIP Rules of the Road," May 2011b.

U.S. Department of Homeland Security

Ryan Consaul and Michael Simpson

DHS, established in 2002, protects the U.S. homeland through a broad array of missions, including border, transportation, and maritime security; cyber and infrastructure security; management of the immigration system and immigration enforcement; disaster response; protection of the nation's leaders; and countering weapons of mass destruction.[1] Because of the unpredictable nature of these missions, DHS must be able to react to changing conditions.

Prior to the formation of DHS, 22 federal departments and agencies existed to perform homeland security–type missions.[2] As shown in Figure 2.1, DHS contains eight operational components (blue) and seven support offices and directorates (green). Recent major organizational changes created CISA as an operational component and the CWMD office as a support component.

DHS is the third-largest federal cabinet-level agency, with more than 250,000 employees and nearly $100 billion in the FY 2023 President's Budget request.[3] The annual DHS budget includes discretionary funding for emerging disasters (the Disaster Relief Fund), which has historically accounted for between 20 and 60 percent of the total budget; discretionary fees; and mandatory appropriations, such as personnel, compensation, and benefits. DHS's budget also includes billions of dollars for major acquisition programs in support of its critical missions. These acquisition programs include a variety of systems to secure the border, secure cyberspace, screen travelers, and conduct other activities.[4]

---

[1] For more information, see DHS, undated-a.

[2] DHS, 2022b.

[3] DHS, 2022b, p. 1.

[4] See GAO, 2022. Because of the significant life-cycle costs associated with major acquisition programs, DHS has implemented policies to oversee the management of these programs to help ensure that they meet cost, schedule, and performance requirements, and DHS maintains mechanisms, such as acquisition program health assessments, to track these areas. Prior to becoming acquisition programs, they must undergo a DHS process to review and validate their requirements to identify synergies across the department and help ensure that acquisition programs fill gaps in capabilities to accomplish DHS's mission. DHS also tracks the extent to which capability gaps have been filled.

FIGURE 2.1
DHS Organizations

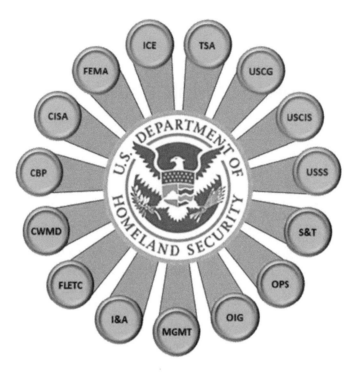

SOURCE: Reproduced from DHS, 2022b.
NOTES: Organizations in blue are DHS components, and organizations in green are support offices or directorates. CBP = U.S. Customs and Border Protection; CISA = Cybersecurity and Infrastructure Security Agency; FEMA = Federal Emergency Management Agency; ICE = U.S. Immigration and Customs Enforcement; TSA = Transportation Security Administration; USCG = U.S. Coast Guard; USCIS = U.S. Citizenship and Immigration Services; USSS = U.S. Secret Service; CWMD = Countering Weapons of Mass Destruction Office; FLETC = Federal Law Enforcement Training Center; I&A = Office of Intelligence and Analysis; MGMT = Management Directorate; OIG = Office of Inspector General; OPS = Office of Operations Coordination (now known as the Office of Homeland Security Situational Awareness, or OSA); S&T = Science and Technology Directorate.

The department's budget submission is organized by programs, projects, and activities (PPAs). Each DHS component's appropriation for non–fee-funded programs is generally divided into four categories:

- operations and support (O&S)
- procurement, construction, and improvements (PC&I)
- research and development (R&D)
- federal assistance (FA).

Like most other U.S. federal agencies, DHS follows an obligation-disbursement approach, in which funds must be obligated before they can be disbursed. This approach also applies to such discretionary funds as the Disaster Relief Fund, which receives *no-year* appropriations that do not expire until they are expended. In addition to the base budget, DHS often receives supplemental funds for emergent requirements, the amount of which varies from year to year, as shown in Figure 2.2.

Prior to 2002, each of the original 22 national security–related organizations had its own budget. When DHS was formed, Congress provided it with budget flexibilities, such as through no-year accounts, to provide it latitude as it was stood up and was allowed to mature.[5] However, DHS faced significant challenges with its financial management because it inherited an array of legacy component systems and processes.[6] Additionally, DHS leadership under former secretaries Tom Ridge, Michael Chertoff, and Janet Napolitano recognized

FIGURE 2.2

DHS's Total Discretionary Budget

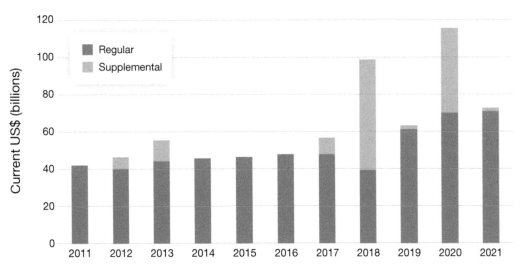

SOURCES: Authors' analysis of data from William L. Painter, *Department of Homeland Security Appropriations: FY2021*, Congressional Research Service, R46802, May 24, 2021; William L. Painter, *Department of Homeland Security Appropriations: FY2022*, Congressional Research Service, R47005, March 24, 2022; OMB, "Historical Tables," White House, undated, Table 5.2— Budget Authority by Agency: 1976–2028; William L. Painter, *DHS Budget v. DHS Appropriations: Fact Sheet*, Congressional Research Service, R44052, April 17, 2019.

NOTE: *Regular* budget authority includes base discretionary appropriations, fee-funded discretionary appropriations, and funding outside the normal appropriations process. It does not include mandatory appropriations or supplemental appropriations. *Supplemental* budget authority consists primarily of funding allocated to the Disaster Relief Fund through supplemental appropriations.

---

[5] DHS officials, interviews with the authors, September–November 2022.

[6] See, for example, Williams, 2004.

a need to increase cohesion within the department and implemented an initiative known as "One DHS."[7]

Under former Secretary Jeh Johnson (2013–2017), DHS sought to institutionalize decision-making bodies and add requirements processes in a "unity-of-effort initiative," thereby seeking to develop the department's PPBE system.[8] However, DHS's federated model, in which the components remain responsible for their separate missions and receive direct appropriations, makes it difficult to promote departmental priorities, unlike in DoD where the Secretary of Defense has more control over the military services and defense agencies. DHS headquarters also lacks the resourcing of OSD, which limits its ability to conduct coordination and management.

Because of a rapidly evolving threat landscape and a politically charged environment, DHS budgeting focuses largely on short-term challenges, and the department's strategic planning capabilities remain immature.[9] However, the department has obtained a clean audit opinion for ten consecutive years,[10] and it recently implemented a consolidated, department-wide PPBE system to harmonize information into a single system of record.[11] DoD could learn from DHS's simplified strategic planning approach, its evaluation efforts, and its consolidated PPBE data system.

Overview of DHS's Budgeting Process

DHS uses a PPBE process that is similar to DoD's to develop a five-year funding plan, known as the Future Years Homeland Security Program (FYHSP) and to inform resource allocation. The PPBE process includes several key annual outputs (listed with tentative time frames based on our discussions with DHS officials):[12]

- Resource planning guidance (RPG) from the DHS Secretary sets departmental priorities and direction (October).
- Fiscal guidance from the DHS Deputy Secretary—through the chief financial officer (CFO)—provides top-line funding allocations to DHS organizations, including the eight operational components (e.g., CBP, ICE, TSA) and support offices (February).

---

[7] See Painter et al., 2019.

[8] For more information, see Johnson, 2014.

[9] DHS official, interview with the authors, September 2022. Also see GAO, 2018; and GAO, 2016a.

[10] Obtaining a *clean opinion* means that DHS has presented its financial statements fairly and in accordance with generally accepted accounting principles. See GAO, 2021; and DHS OIG, 2022.

[11] DHS official, interview with the authors, October 2022.

[12] DHS Instruction 101-01-001, *Planning, Programming, Budgeting, and Execution*, U.S. Department of Homeland Security, June 11, 2019.

- Resource allocation plans (RAPs) developed by the DHS operational components and support offices propose five-year organizational funding plans (April).
- The DHS Secretary issues a resource allocation decision (RAD) (July/August), which is then negotiated with OMB (September/October).
- The FYHSP and the President's Budget request are sent to Congress (February).
- DHS conducts a strategic review process to self-assess implementation progress (no time frame).

The first five outputs cover just 15 months of the flowchart shown in Figure 2.3 (the blue cycle and the first half of the orange cycle, which are the programming and budgeting phases of the PPBE process). However, in each calendar year, multiple cycles of PPBE and reporting are underway at any given time, as illustrated in the figure.

DHS has been required to use a PPBE system almost since the department's inception, but the system's use has evolved over time. Section 874 of the 2002 Homeland Security Act—DHS's enabling legislation—required the department to "develop a Future Years Homeland Security Program," modeled after DoD's FYDP, to anticipate DHS's five-year future budget

FIGURE 2.3

Flowchart of the DHS Budgeting Process

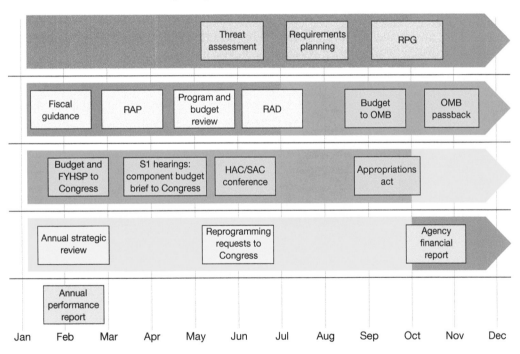

SOURCE: Adapted from DHS Office of the Chief Financial Officer Program Analysis and Evaluation, briefing provided to the authors, August 2023.
NOTE: This figure shows that multiple PPBE cycles are underway at any one time. Phases are color coded from top to bottom as follows: planning (gray), programming (blue), budgeting (orange), execution (yellow), and reporting (green). HAC = House Appropriations Committee; S1 = DHS Secretary; SAC = Senate Appropriations Committee.

requests and to "accompany its budget submissions."[13] Congress amended the Homeland Security Act in 2005 to formally mandate the use of a PPBE system. In particular, the 2005 amendments established the DHS Office of Program Analysis and Evaluation (PA&E), which would report to the department's CFO in implementing PPBE within DHS. Congress thus led the push for DHS to adopt a program-based budgeting approach as a means both to fulfill the department's performance measurement and program evaluation requirements under the Government Performance and Results Act (GPRA) of 1993 and to mimic DoD's process for tying strategic objectives to program budgets.[14]

DHS structured its programs around its strategic goals, implicitly linking plans to programs.[15] Congress's stated intent was for DHS to integrate its programming function with the strategic plan and annual planning cycles. However, a combination of CFO and PA&E leadership changes, a lack of analytic capacity within the components, and a lack of senior leader buy-in left DHS's implementation of its PPBE system focused primarily on budgeting and execution—not planning or programming—for the first several years of the department's existence.[16]

It was not until 2010 that DHS began implementing a formal policy guidance process meant to inform the component RAPs. After Secretary Jeh Johnson issued his "Strengthening Departmental Unity of Effort" memorandum in 2014, DHS began developing its RPG.[17] The Deputy's Management Action Group (DMAG), chaired by the DHS Deputy Secretary, then began performing a prioritization drill in which it identified key priorities from the DHS Strategic Plan that should inform the components' planning and programming processes.

Decisionmakers and Stakeholders

The DHS PPBE process is driven by a mix of headquarters and component decisionmaking. According to DHS Directive 101-01, the DHS Secretary is the ultimate decision authority, provides direction through DHS strategic guidance and the RPG, and is the deciding authority for the RAD.[18] The Deputy Assistant Secretary for Strategic Integration in the Office of Strategy, Policy, and Plans coordinates inputs and manages the planning phase, which culminates in the issuance of the RPG. The CFO provides DHS components with fiscal guidance. The director of PA&E in the Office of the CFO manages the PPBE process

---

[13] William F. West, *Program Budgeting and the Performance Movement: The Elusive Quest for Efficiency in Government*, Georgetown University Press, 2011, p. 139.

[14] Public Law 103-62, Government Performance and Results Act of 1993, August 3, 1993.

[15] West, 2011, p. 140.

[16] West, 2011, p. 140.

[17] DHS official, interview with the author, September 2022.

[18] DHS Directive 101-01, *Planning, Programming, Budgeting, and Execution*, rev. 1, U.S. Department of Homeland Security, June 4, 2019.

and integrates it with other DHS governance processes.[19] The DMAG provides resource allocation recommendations to the Senior Leaders Council (SLC), which comprises the DHS Secretary and component heads.[20]

These forums are intended to provide senior decisionmakers across DHS with a say in the PPBE process and are composed of both political appointees and civil servants. However, these governance bodies have not been institutionalized, and DHS leadership may use them in different ways as it sees fit. For example, under the Trump administration, these governance bodies were not as active as in prior years, according to DHS officials.[21] In addition, component input is provided up through the RAPs that inform the RAD. As with other federal departments, Congress has no role in building DHS's budget, but Congress is central in reviewing the department's budget request and making appropriations that may or may not align with its budget.

Planning and Programming

There is significant unpredictability in DHS's missions, which affects the department's ability to conduct strategic planning. The planning phase culminates in the RPG issued by the DHS Secretary.

Preceding the planning phase, the DHS Office of Intelligence and Analysis has historically conducted a threat assessment to inform the department's planning and strategy development.[22] The DHS Office of Strategy, Policy, and Plans coordinates the planning phase and obtains input from stakeholders across the department, including the components.[23] In past years, the office attempted to prioritize across DHS missions to inform its budget formulation, but this effort was viewed as unsuccessful and was discontinued because DHS leadership could not prioritize across the varied missions; essentially, everything became a priority.[24] The office also conducted winter studies that were intended to be analytic inputs to the planning phase, but these were also discontinued because of component disengagement and inability to inform the process.[25] DHS did not release its 2018 Quadrennial Homeland Security Review (QHSR), which would have provided strategic guidance for the homeland security enterprise. Although the RPG could not leverage a 2018 QHSR, DHS officials we spoke with said that prior leadership had found these reviews to be of limited value anyway, so it is

---

[19] DHS Instruction 101-01-001, 2019, p. 8.

[20] DHS officials told us that the SLC rarely met, and the DMAG usually provided recommendations to the Secretary of Homeland Security for consideration (DHS officials, interviews with the authors, September 2022).

[21] DHS officials, interview with the authors, September 2022.

[22] Former DHS official, interview with the authors, January 2023.

[23] DHS Instruction 101-01-001, 2019, p. 8.

[24] DHS official, interview with the authors, September 2022.

[25] DHS official, interview with the authors, September 2022.

unclear whether the QHSR would have influenced the RPG.[26] As a result, the planning phase remains immature and in a state of transition as of this writing. DHS is looking to develop more-robust analytic inputs to inform the RPG and conduct more in-depth analysis following the creation of the budget to better understand how it reflects the RPG and what trade-offs were made to ensure that it would do so.[27]

The programming phase centers on the development and approval of the RAPs. Components manage this process internally. Individual offices within the components develop submissions to inform the RAPs based on their future resource needs. The RAPs are then reviewed, culminating in each component's RAP submission to DHS headquarters, which reviews the RAPs to determine their alignment with the RPG and the Secretary's priorities.[28] The programming process concludes with the creation of the RAD. The RAD is used to develop the FYHSP and congressional budget request, which is developed in the budgeting phase.

Budgeting and Execution

The budgeting phase overlaps with the programming phase in a few key ways. The budgeting phase is overseen by the DHS CFO. DHS Instruction 101-01-001 delegates authority for the budgeting phase to the DHS budget director within the Budget Division of the CFO's office. Although the PPBE process ultimately produces the five-year FYHSP, the budgeting phase is concerned primarily with developing a "fully justified one-year budget submission" that results in a congressional appropriation.[29]

The budgeting phase begins after DHS components have submitted their RAPs, enabling the Budget Division to inform the RAD. Budget formulation is at once distinct from longer-term PPBE resource decisions, with a primary focus on the annual budget cycle, and an important input for RADs and the five-year FYHSP. The Budget Division leads the budget review, which includes key pricing and execution considerations. The division is also responsible for preparing rigorous budget justifications and supporting materials, which are initially submitted to OMB and subsequently prepared for Congress.[30]

To aid in budget formulation, DHS began implementing a new consolidated budget system, known as the PPBE One Number system, in 2020. DHS had identified inefficiencies in its past systems and sought to replace outdated tools with a modernized system that would provide the department with advanced capabilities to manage the development

---

[26] DHS officials, interviews with the authors, September–October 2022.

[27] DHS official, interview with the authors, October 2022.

[28] DHS Instruction 101-01-001, 2019.

[29] DHS Directive 101-01, 2019, p. 6.

[30] DHS Instruction 101-01-001, 2019.

and execution of its budget.[31] The PPBE One Number system sought to consolidate several tools: the FYHSP programming tool, which captured FYHSP-level program, performance, and mission data; the Budget Formulation Execution Manager, which was a web-based application used to create budget justification documents; and numerous Microsoft Excel database–capture tools.[32]

The PPBE One Number system offers DHS a consolidated tool for budget formulation, performance management, and monthly obligation planning while eliminating disparate tools and the need to reenter data into multiple systems and spreadsheets. The system launched in July 2020 with a performance management workflow and was expanded in August 2020 with budget formulation capabilities. In January 2021, the system went live with capabilities for monthly obligation planning, and preliminary evidence suggests that it has helped reduce duplicative reentry of data, along with allowing users to share data more easily and automating previously manual reports.[33]

As noted, DHS components submit budget estimates for review by the Budget Division, which assists in the preparation of congressional budget justifications. Although many components evaluate priorities in their own capability portfolios during the programming phase, the final budget submission is organized by PPAs across the four appropriation categories described earlier.[34]

DHS funds are typically budgeted annually, but some programs receive multiyear or no-year appropriations. For example, Congress sometimes appropriates multiyear funds to major acquisition programs to foster a stable production and contracting environment. In contrast, a key example of no-year money is the Disaster Relief Fund, which is meant to provide FEMA with the flexibility to respond quickly to emerging disaster relief and recovery needs. In another example, DHS officials mentioned how the border security, fencing, infrastructure, and technology appropriation provided DHS the ability to carry over significant amounts of funding related to this mission area.[35] However, they noted that funds are no longer appropriated to this account. DHS officials stated that the use of no-year appropriations was significantly curtailed with the implementation of the common appropriations structure.

As for the duration of funding availability, the House report accompanying the FY 2019 DHS appropriations bill outlined different periods of availability for three of the four major account types: "With limited exception, the O&S accounts shall have one year of availability; the PC&I accounts shall have five years of availability for construction and three years for all

---

[31] DHS official, interview with the authors, October 2022.

[32] DHS official, interview with the authors, October 2022.

[33] DHS official, interview with the authors, October 2022.

[34] The four appropriation categories are O&S, PC&I, R&D, and FA.

[35] DHS officials, interview with the authors, November 2022. For more information on the border security, fencing, infrastructure, and technology appropriation, see William L. Painter and Audrey Singer, *DHS Border Barrier Funding*, Congressional Research Service, R45888, January 29, 2020.

other activities; and the R&D accounts shall have two years of availability."[36] However, Congress authorized DHS components to carry a portion of their one-year O&S accounts forward into the next fiscal year and execute up to 50 percent of their prior-year lapsed (unobligated) balance amounts. This authorization is worded as follows in annual appropriations bills:

> Section 505. The Committee continues and modifies a provision providing that not to exceed 50 percent of unobligated balances from prior year appropriations for each Operations and Support appropriation shall remain available through fiscal year 2020, subject to section 503 reprogramming requirements.[37]

This authority was still in effect as of this writing. However, this authority does not apply to multiyear O&S accounts or other categories of accounts. According to DHS officials, multiyear O&S accounts are rarer than one-year O&S accounts, which are typically related to personnel compensation.[38] Seven DHS components employ fee-and-fine authorities that may "offset appropriations" or "cover the cost of provided services."[39] Budget submissions thus articulate these additional account categories, along with mandatory appropriations.[40]

Whereas the early PPBE phases are assigned a single decision authority (the DHS Secretary), the execution phase is "owned by many parties in DHS including senior leadership, the DHS CFO and Performance Improvement Officer, and Components."[41] Ultimately, however, responsibility for disbursing and monitoring obligated funds falls to the components.[42] Although the Budget Division in the CFO's office monitors obligation rates and oversees department-wide requests for information regarding execution, the components are ultimately responsible for monitoring outlays and expenditure rates.[43] According to interviews with DHS officials, the DHS CFO Budget Division is primarily focused on monitoring com-

---

[36] U.S. House of Representatives, Committee on Appropriations, *Department of Homeland Security Appropriations Bill, 2019*, House Report 115-948, U.S. Government Publishing Office, September 12, 2018.

[37] U.S. House of Representatives, Committee on Appropriations, 2018.

[38] DHS officials, interview with the authors, November 2022.

[39] See DHS, Office of the Chief Financial Officer, *A Common Appropriations Structure for DHS: FY 2016 Crosswalk*, Addendum to the Fiscal Year 2016 President's Budget, February 2, 2015, p. 3. These components are as follows: CBP; U.S. Citizenship and Immigration Services; TSA; FEMA; ICE; National Protection and Programs Directorate, which has been reorganized as CISA; and USCG. Examples of fee-and-fine authorities include fees to recover the cost to secure federal facilities, offset aviation security costs, and process immigration benefit requests. For further reading on DHS's fee-based programs, see GAO, *DHS Management: Enhanced Oversight Could Better Ensure Programs Receiving Fees and Other Collections Use Funds Efficiently*, GAO-16-443, July 21, 2016b.

[40] DHS, *U.S. Department of Homeland Security FY 2020–2022 Annual Performance Report*, undated-b, p. 5.

[41] DHS Instruction 101-01-001, 2019, p. 15.

[42] DHS officials, interview with the authors, September 2022.

[43] DHS officials, interview with the authors, September 2022.

ponent obligations within the execution phase rather than expenditure rates and outlays. The Budget Division approves quarterly apportionment documents before they are sent to OMB.

According to DHS officials, the principal objective of the execution phase, as outlined in both DHS Directive 101-01 and DHS Instruction 101-01-001, is "to responsibly expend resources and to account for cost and performance to determine if value is delivered to our stakeholders."[44] Importantly, analytic and reporting requirements extend to both financial expenditures *and* broader performance measures. The execution phase should also include an assessment of risks and should facilitate continuous process improvement.[45] Finally, the execution phase acts as a feedback mechanism for earlier PPBE phases, and therefore, DHS should apply appropriate measures and metrics to specific PPBE processes and procedures in this phase.[46]

The DHS budget director is responsible for managing the "execution of apportionment and control of funds processes, including transfers and reprogramming."[47] As part of this oversight process, the budget director conducts status-of-funds and midyear reviews to ensure that funds are being executed in accordance with congressional intent. During the execution phase, programs may require additional support to cover emerging needs or unforeseen crises. DHS must initiate a reprogramming request to Congress before June 30 if the required additional support exceeds 10 percent of its originally appropriated funding.[48] If current-year appropriations are passed under a continuing resolution, the DHS *Financial Management Policy Manual* indicates that, generally, no new PPAs, new hiring, or new contract awards may occur.[49] Rather, DHS must work within the prior fiscal year account structure.

DHS PA&E is charged with developing metrics for program assessment during the planning and programming phases, in accordance with the GPRA. PA&E does not have a formal role in the execution phase, but DHS components coordinate with PA&E on performance assessments; quarterly reports detail progress toward performance goals.[50]

---

[44] DHS officials, interview with the authors, September 2022.

[45] DHS Instruction 101-01-001, 2019, p. 6. DHS is to conduct a strategic review in the execution phase. DHS notes that the strategic review serves multiple purposes, such as conducting a self-assessment on progress and ways to improve; facilitating organizational best practices; making findings available to senior leaders to inform planning, budgeting, and management decisions; and informing conversations with OMB to inform its activities (DHS Instruction 101-01-001, 2019). However, it is unclear how the information may specifically drive budget decisions.

[46] DHS Instruction 101-01-001, 2019, p. 15.

[47] DHS Instruction 101-01-001, 2019, p. 10.

[48] Michelle Mrdeza, and Kenneth Gold, "Reprogramming Funds: Understanding the Appropriators' Perspective," Government Affairs Institute at Georgetown University, undated.

[49] DHS, *Financial Management Policy Manual*, September 8, 2020, Not available to the general public. Last issued in September 2021, this manual is the primary source for departmental financial management policies and covers a wide variety of topics. It is organized into separate chapters and sections within chapters, and, based on our estimation, it is approximately 1,400 total pages.

[50] DHS Instruction 101-01-001, 2019, p. 19; DHS officials, interview with the authors, September 2022.

The DHS evaluation officer oversees the department's evaluation activities and its compliance with the evaluation elements required by the Foundations for Evidence-Based Policymaking Act of 2018 (also known as the Evidence Act).[51] In particular, the evaluation officer coordinates an Annual Evaluation Plan, as well as assessments of the department's evaluation capabilities every four years—resulting in a four-year "learning agenda" that ties long-term strategic priorities to performance evaluation criteria.[52] Each program in the DHS components likewise has an assessment team that reviews how effectively the program's mission priorities are executed. Some components centralize program evaluation and performance management functions within the CFO organization.[53] Ultimately, component strategic and performance assessments are intended to inform the following year's annual budget.

Oversight

DHS's PPBE process takes almost two years, from the beginning of the planning phase to the delivery of the budget request to Congress. Along the way, multiple reviews by DHS leadership and OMB provide opportunities to adapt the budget to the changing homeland security–related landscape and priorities.

Once the President's Budget is released (which by law is supposed to occur in early February), the House and Senate homeland security authorization and appropriations committees begin reviewing it. This review occurs through staff briefings with DHS and component officials and culminates in committee hearings. The House and Senate committees conduct budget hearings with the DHS Secretary each year. The appropriations committees hold additional hearings with the components and support offices. Their subcommittees then typically release the text of the DHS appropriations bill in the summer and add subcommittee and full committee markups. For full committee markups, the committees attach accompanying reports to the bill, which can provide additional guidance. With the exception of the House-passed FY 2024 DHS appropriations bill, homeland security appropriations bills had not received individual floor consideration by either the House or the Senate in the previous five years. Rather, homeland security appropriations are routinely passed via continuing resolutions or omnibus (i.e., consolidated) appropriations bills. Furthermore, DHS's authorizing legislation, the Homeland Security Act of 2002, has not received a comprehensive reauthorization since its enactment.

Each appropriations bill outlines the authorities for transfers and reprogramming by DHS. The following quoted passage notes the Section 502 and 503 authorities that were included in the FY 2022 DHS appropriations bill. Although there are certain exceptions that exist

---

[51] Public Law 115-435, Foundations for Evidence-Based Policymaking Act of 2018, January 14, 2019.

[52] DHS Instruction 069-03-001, *Program, Policy, and Organizational Evaluations*, U.S. Department of Homeland Security, February 16, 2021, p. 5.

[53] DHS Instruction 101-01-001, 2019, p. 19.

throughout the bill, the following authorities remain in effect during a continuing resolution until the enactment of FY 2023 appropriations:[54]

SEC. 502. Subject to the requirements of section 503 of this Act, the unexpended balances of prior appropriations provided for activities in this Act may be transferred to appropriation accounts for such activities established pursuant to this Act, may be merged with funds in the applicable established accounts, and thereafter may be accounted for as one fund for the same time period as originally enacted.

SEC. 503. (a) None of the funds provided by this Act, provided by previous appropriations Acts to the components in or transferred to the Department of Homeland Security that remain available for obligation or expenditure in fiscal year 2022, or provided from any accounts in the Treasury of the United States derived by the collection of fees available to the components funded by this Act, shall be available for obligation or expenditure through a reprogramming of funds that—

(1) creates or eliminates a program, project, or activity, or increases funds for any program, project, or activity for which funds have been denied or restricted by the Congress;

(2) contracts out any function or activity presently performed by Federal employees or any new function or activity proposed to be performed by Federal employees in the President's budget proposal for fiscal year 2022 for the Department of Homeland Security;

(3) augments funding for existing programs, projects, or activities in excess of $5,000,000 or 10 percent, whichever is less;

(4) reduces funding for any program, project, or activity, or numbers of personnel, by 10 percent or more; or

(5) results from any general savings from a reduction in personnel that would result in a change in funding levels for programs, projects, or activities as approved by the Congress.

(b) Subsection (a) shall not apply if the Committees on Appropriations of the Senate and the House of Representatives are notified at least 15 days in advance of such reprogramming.

(c) Up to 5 percent of any appropriation made available for the current fiscal year for the Department of Homeland Security by this Act or provided by previous appropriations Acts may be transferred between such appropriations if the Committees on Appropriations of the Senate and the House of Representatives are notified at least 30 days in advance of such transfer, but no such appropriation, except as otherwise specifically provided, shall be increased by more than 10 percent by such transfer.

---

[54] The block quotation reproduces Public Law 117-103, Consolidated Appropriations Act, 2022, Division F, Title V, General Provisions, March 15, 2022.

(d) Notwithstanding subsections (a), (b), and (c), no funds shall be reprogrammed within or transferred between appropriations based upon an initial notification provided after June 30, except in extraordinary circumstances that imminently threaten the safety of human life or the protection of property.

(e) The notification thresholds and procedures set forth in subsections (a), (b), (c), and (d) shall apply to any use of deobligated balances of funds provided in previous Department of Homeland Security Appropriations Acts that remain available for obligation in the current year.

(f) Notwithstanding subsection (c), the Secretary of Homeland Security may transfer to the fund established by 8 U.S.C. 1101 note, up to $20,000,000 from appropriations available to the Department of Homeland Security: Provided, That the Secretary shall notify the Committees on Appropriations of the Senate and the House of Representatives at least 5 days in advance of such transfer.

In addition to receiving funds from Congress, DHS receives equipment transfers and loans from DoD. For example, DoD transferred tactical aerostat systems used for surveillance to CBP after their use in Afghanistan.[55] CBP and the USCG have relied on H-60 helicopters, and most of those operated by CBP were on loan from the U.S. Army.[56] These examples demonstrate how DHS can rely on DoD systems to support its critical missions.

As described in the Financial Management Policy Manual, DHS has monitoring mechanisms in place, such as obligation plans, expenditure plans, and budget execution plans.[57] DHS receives an independent audit of its financial statements annually, contracted out by DHS OIG. Accounting firm KPMG LLP performed the audit for FYs 2021 and 2022.[58] For ten consecutive years, DHS has received an unmodified (i.e., *clean*) opinion on its financial statements. Although such an opinion does not guarantee that funds are spent effectively, it does indicate that financial records are accurate, and that, in turn, can enable more-effective oversight.

Analysis of DHS's Budgeting Process

Strengths

We identified five strengths in DHS's budget process. First, the department's PPBE process offers some level of *transparency* in decisionmaking. In each phase, guidance—including the RPG, fiscal guidance, and the RAD—is documented and disseminated across DHS.

---

[55] Dave Long, "CBP's Eyes in the Sky," *Frontline Magazine*, U.S. Customs and Border Protection, last updated April 11, 2016.

[56] DHS OIG, *DHS' H-60 Helicopter Programs*, OIG-13-89, revised May 2013.

[57] DHS, 2020.

[58] DHS OIG, 2022.

Components document their decisions through RAP submissions. The SLC and DMAG provide forums for senior leaders to make resource decisions. These mechanisms, to the extent that they are fully used, provide senior leaders with visibility into the department's resourcing decisions.

Second, the director of PA&E in the Office of the CFO acts as the *single owner* responsible for managing the PPBE process across DHS. Having a single process owner helps integrate all key stakeholders and reduces the risk of gaps between phases. It also helps standardize language, methods, and culture across PPBE processes by situating ownership within the Office of the CFO.

The third strength has been the migration to a *common appropriations structure*. Beginning with the FY 2017 budget request, DHS, in coordination with Congress, consolidated most of its appropriation accounts into four categories.[59] Doing so allowed DHS to reduce confusion, increase consistency across the components, and compare its accounts.[60] Eliminating legacy account structures and reexamining existing budget approaches were important steps in the department's maturation and unification.

Fourth, DHS has had *flexibility* in carrying over significant amounts of funds for certain accounts, such as an account for border security fencing, infrastructure, and technology, as well as unobligated O&S funds. These examples show how DHS has flexibility to meet evolving mission needs.

Fifth, *consolidating budget tools* into the PPBE One Number System has strengthened DHS's budget formulation and execution processes. Despite challenges during its development, the system has allowed budget officials to share and analyze budget data more easily. It assists with oversight requests by reducing the burden of manual data gathering. The modernized system also provides an opportunity to make updates and upgrades as the needs of DHS's budget staff evolve.

Challenges

We identified four challenges in DHS's budget process. First, the *planning phase lacks strategic inputs and consistent supporting analysis*. The department did not release its most recently mandated QHSR in 2018. In the absence of this document, the DHS Strategic Plan for FYs 2020–2024 was the most recent strategic document that could inform the planning phase.[61] Previously, DHS had conducted short-term analytic efforts, known as winter studies, on cross-component and departmental issues of interest. However, those studies were intended to inform resource planning efforts, and they were seen as uninformative and discontinued. The DHS Secretary can also create issue teams to examine certain topics, but those teams and initiatives could change from year to year with changing priorities.

---

[59] Painter, 2021.

[60] DHS officials, interview with the authors, September 2022.

[61] DHS, undated-a.

Second, DHS has exhibited a *structural inability to set priorities*. The department has attempted to prioritize issues and programs across its missions. These efforts have not been viewed favorably by some internal stakeholders. DHS has faced challenges prioritizing its missions and programs because they are too varied. The department's federated model, in which components remain responsible for specific missions and receive direct appropriations, makes it difficult to promote departmental priorities. DHS lacks a Goldwater-Nichols imperative to compel jointness, and DHS headquarters lacks the resources of OSD to do so. Given the unpredictability of DHS's mission sets, it has been difficult to conduct long-term strategic planning and to foster DHS's strategic planning capabilities.

Third, *the SLC and DMAG have not been institutionalized*, meaning that DHS leadership can choose whether to use or ignore these functions. These bodies, first established under Secretary Jeh Johnson, were intended to inform leadership decisions, including those related to resources. Under the subsequent administration, the SLC and DMAG met less frequently. By using the SLC and DMAG inconsistently, DHS leadership adds unpredictability to the PPBE process, making it more difficult to implement and, subsequently, to make transparent resource decisions.

Fourth, *DHS headquarters has limited visibility into expenditure and outlay rates,* which are recorded and monitored by the components. Each component employs a unique data system to track expenditures, which precludes an integrated department-wide assessment of execution following the obligation of funds. This lack of centralized authority over expenditures potentially hinders DHS's visibility and control over the execution phase.[62]

Applicability

DHS's budget process is generally similar to that of DoD, with some exceptions, simplifications, and less formalization. Whereas DHS and DoD both issue planning guidance, DHS is not statutorily required to do so. DHS guidance is intended to articulate the DHS Secretary's priorities to be reflected in the component RAPs, but the process of developing the DHS Secretary's RPG remains in flux and immature as of this writing.

DHS does not prioritize across missions, unlike DoD combatant commanders, who prioritize across missions by developing integrated priority lists. Because DHS lacks an organization with OSD- or Joint Staff–like functions, DHS also does not conduct program assessments akin to those performed by the Joint Staff. The DHS Joint Requirements Council *should* validate DHS program requirements; however, despite having budgeted programs, the components' requirement prioritization processes vary in their maturity.

Related to oversight, DHS does not receive routine congressional authorizations. Although legislation has authorized certain components and programs, there has not been

---

[62] A DHS official stated that DHS's PPBE One Number System imports execution data to inform exhibits for budget justification documentation (DHS official, interview with the authors, October 2022).

a comprehensive authorization comparable to the annual NDAA since the Homeland Security Act of 2002.

Lessons from DHS's Budgeting Process

Lesson 1: DHS Has a More Flexible, If Less Stable, Planning Process Than DoD

DHS has a simpler strategic planning process than DoD. It is predicated on clear and transparent strategic direction from the DHS Secretary. The DHS planning process relies on fewer inputs and reviews and is, therefore, less robust than DoD's process. DHS does not have requirements for the planning phase codified in statute or policy, which gives DHS latitude to adapt its planning process. Although this gives DHS some flexibility, there are important trade-offs in terms of stability, predictability, and adherence to strategy that mature strategic planning processes reflect. DoD should not emulate DHS's planning process but could benefit from a review of its process inputs and review processes to determine any potential efficiencies, so long as the potential changes do not harm the robustness of DoD's PPBE process.

Lesson 2: DHS Has Invested in Evaluations

To support implementation of the Evidence Act, DHS issued a policy in February 2021 on program, policy, and organizational evaluations. It also developed annual evaluation plans for FYs 2022 and 2023. This line of effort demonstrates an investment by DHS in evaluation. DHS's efforts in this area could help inform DoD's approach to the execution phase.

Lesson 3: DHS Has Benefited from a Consolidated PPBE System

DHS's consolidation of its PPBE information system has enhanced its ability to create and manage budgets. The consolidation has combined the systems for generating congressional budget justification documents, developing the FYHSP, and capturing performance management data. DHS officials have told us that implementing a consolidated system has reduced their reliance on Microsoft Excel spreadsheet templates and data reentry, and it has automated the generation of certain reports that were previously created manually. DoD should examine the feasibility of implementing such a consolidated system and whether the benefits of doing so would outweigh the costs.

Table 2.1 summarizes these three lessons.

TABLE 2.1

Summary of Lessons from DHS's Budgeting Process

Theme	Lesson Learned	Description
Planning and programming	Lesson 1: DHS has a more flexible, if less stable, planning process than DoD.	DoD should not emulate the DHS process but should assess the necessity of DoD planning phase inputs and planning phase review processes.
Budgeting and execution	Lesson 2: DHS has invested in evaluations.	DHS efforts to strengthen evaluation could help inform DoD approaches to execution.
Oversight	Lesson 3: DHS has benefited from a consolidated PPBE system.	DoD should explore the feasibility and the costs and benefits of implementing a consolidated PPBE information system.

CHAPTER 3

U.S. Department of Health and Human Services

Michael Simpson and Devon Hill

HHS, established with its current organizational structure and mission in 1980, provides health and human services and pursues scientific advances in the areas of medicine, public health, and social services. The department has five primary strategic goals: protecting and strengthening equitable access to high-quality and affordable health care; safeguarding and improving national and global health outcomes; strengthening social well-being, equity, and economic resilience; restoring trust and accelerating progress in science and research; and advancing strategic management to build trust, transparency, and accountability.[1]

HHS traces its roots as a federal agency to the creation of the Federal Security Agency in 1939 and the subsequent elevation of HEW to cabinet-level status in 1953.[2] Since then, the department has undergone several major reorganizations, including the U.S. Department of Education's elevation to a stand-alone, cabinet-level department in 1979; HEW's renaming to the U.S. Department of Health and Human Services in 1980; and the Social Security Administration's spin-off as an independent agency in 1994. As shown in Table 3.1, HHS has 12 OPDIVs and 14 Office of the Secretary staff divisions. The OPDIVs are responsible for administering HHS's Public Health Service missions,[3] such as protecting the U.S. public from pathogens and other public health threats, conducting health and biomedical research, and promoting the economic and social well-being of U.S. communities. Staff divisions provide

---

[1] These goals are set by the HHS Secretary at the start of a new administration. Therefore, the strategic goals guiding the department are likely to vary from administration to administration. For more information, see HHS, 2022a.

[2] HHS, 2023a.

[3] Nine of the 12 OPDIVs formally constitute the U.S. Public Health Service agencies. These are as follows: NIH, FDA, CDC, Health Resources and Services Administration, Agency for Toxic Substances and Disease Registry, Indian Health Service, Agency for Healthcare Research and Quality, SAMHSA, and Administration for Strategic Preparedness and Response.

TABLE 3.1

HHS Organizations

Operating Divisions	Staff Divisions
• Administration for Children and Families • Administration for Community Living • Agency for Healthcare Research and Quality • Administration for Strategic Preparedness and Response • Agency for Toxic Substances and Disease Registry • CDC • Centers for Medicare and Medicaid Services • FDA • Health Resources and Services Administration • Indian Health Service • NIH • SAMHSA	• ASA • ASFR • Office of the Assistant Secretary for Health • Office of the Assistant Secretary for Legislation • ASPE • Office of the Assistant Secretary for Public Affairs • Office for Civil Rights • Departmental Appeals Board • Office of the General Counsel • Office of Global Affairs • OIG • Office of Medicare Hearings and Appeals • Office of the National Coordinator for Health Information Technology • Chief Information Officer

SOURCE: Features information from HHS, 2023b.

support to the Office of the Secretary of Health and Human Services in administering all HHS programs and activities.[4]

A decade after its 1953 elevation to a cabinet-level department, HEW adopted an early version of PPBE as part of the Johnson administration's effort in 1965 to establish processes for program budgeting in all U.S. federal civilian agencies. HEW modeled its program budgeting framework and organizational structure after DoD's PPBE model, which included creating an office (ASPE) to oversee evaluation and budget trade-offs and linking strategic planning efforts to five-year budget plans.[5] Although some vestiges of this framework—such as its rigorous program evaluation capabilities—remain features of the contemporary HHS budgeting system, the department gradually dismantled much of its earlier PPBS during the 1970s in

---

[4] *Programs* and *activities* have specific budgetary definitions and implications as established by OMB Circular No. A-11, *Preparation, Submission, and Execution of the Budget*, Executive Office of the President, August 2022. In particular, *programs* are elements within an agency's budget accounts that contain a variety of projects with common purposes, functions, or goals. *Activities* refer to two distinct ways of presenting information in budget accounts: budget activities and program activities. *Budget activities* are "specific and distinguishable line[s] of work performed by a governmental unit to discharge a function or subfunction for which the governmental unit is responsible" and are typically an umbrella for categorizing programs with similar functions or purposes within individual budget accounts. In contrast, *program activities* refer to the "operations financed by a specific budget account" and are presented within individual program structures in agency budget documentation. For example, Chronic Disease Prevention and Health Promotion is one budget activity in the CDC's appropriations that covers various individual programs, such as Cancer Prevention and Control and Health Promotion (GAO, *A Glossary of Terms Used in the Federal Budget Process*, GAO-05-734SP, September 2005).

[5] Rivlin, 1969, p. 911.

response to the perception that PPBS did not fit with HEW's missions, organizational structure, or program needs.[6] Despite its shared legacy with DoD's PPBE, HHS's budgeting system has diverged significantly from DoD's since 1980.

Because HHS programs focus on delivering health care services and grants, outside its mandatory funding, the department operates primarily on one-year discretionary funding and restricts budget planning to the annual budget cycle.[7] Consequently, HHS does not engage in robust long-term budget planning, nor does it have well-established links between strategic planning and budgeting.[8] However, the department benefits from considerable budgetary flexibility, which Congress provides so that HHS can respond to unpredictable mission needs. The department also benefits from consistent policies and processes, effective mechanisms for adjudicating budget priorities and trade-offs, relative effectiveness in managing continuing resolutions, and strong transparency and oversight mechanisms. DoD could learn from HHS's collaborative top-down and bottom-up budgeting process, its budget flexibility, and its centralized oversight mechanisms.

Overview of HHS's Budgeting Process

HHS operates the largest budget by far (in terms of total budget authority) among the cabinet-level agencies. Its discretionary budget—$183 billion in 2021—is second only to that of DoD.[9] HHS's total annual budget authority rose to nearly $1.7 trillion in its FY 2021 appropriation, which was more than double DoD's total 2021 appropriation of roughly $800 billion.

Mandatory HHS funding (primarily for Medicare and Medicaid[10]) constitutes about 90 percent of the total HHS budget.[11] Residual HHS budget authority covers discretionary appropriations for ongoing operating and staff division activities, which account for about 8–10 percent of the total budget and for discretionary fees (especially user fees collected from pharmaceutical companies by the FDA).[12] Funding for mandatory programs is managed

---

[6] See, for example, Harlow, 1973, p. 90; Jablonsky and Dirsmith, 1978, p. 216; Rivlin, 1969, p. 922; and U.S. General Accounting Office, 1990, p. 22.

[7] HHS officials, interviews with the authors, October 2022–January 2023.

[8] HHS officials, interviews with the authors, October 2022–January 2023.

[9] OMB, undated, Table 5.2.

[10] The federal government and the states split the cost of funding Medicaid; the federal government matches a proportion of state expenses. Most Medicaid expenses incurred by the states are reimbursed by the federal government according to a calculated federal medical assistance percentage. For further reading on Medicaid financing, see Alison Mitchell, *Medicaid Financing and Expenditures*, Congressional Research Service, R42640, November 10, 2020.

[11] Jessica Tollestrup, Karen E. Lynch, and Ada S. Cornell, *Department of Health and Human Services: FY2023 Budget Request*, Congressional Research Service, R47122, May 31, 2022, p. 4.

[12] Other FDA user fees include accreditation and certification payments.

separately under each program's authorizing statute.[13] Most mandatory programs, such as Medicare and children's entitlement programs, are budgeted on ten-year schedules outside the annual appropriations process, but some so-called discretionary entitlements (notably Medicaid) are subject to annual congressional oversight.[14]

HHS's base discretionary budget includes grants to researchers, state and local governments, and educational and community organizations;[15] all major Public Health Service operations; internal research and program evaluation activities; and basic and applied biomedical research in disease causes and prevention, medical countermeasures, and other areas relevant to Public Health Service missions. In addition to its base budget, HHS often receives supplemental discretionary appropriations to address ongoing public health crises.

The volume of supplemental appropriations varies by year. For instance, supplemental funding for coronavirus disease 2019 (COVID-19) pandemic relief has been the primary driver of historically large increases to the HHS budget since 2020, as can be seen in Figure 3.1. Congress has appropriated more than $400 billion since March 2020 to HHS agencies for "the domestic [COVID-19] public health response," primarily as "emergency-designated supplemental discretionary appropriations."[16] In recent years, Congress has also appropriated

FIGURE 3.1

HHS's Total Budget Authority, 2010–2021

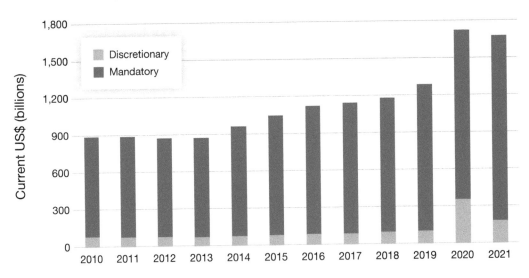

SOURCE: Features information from OMB, undated, Table 5.2.

---

[13] Tollestrup, Lynch, and Cornell, 2022, p. 4.

[14] Tollestrup, Lynch, and Cornell, 2022, p. 4.

[15] HHS is the largest grantmaking authority in the U.S. government.

[16] Some emergency-supplemental funds have been appropriated as mandatory spending, however. See Kavya Sekar, *COVID-19 Supplemental Appropriations for the Department of Health and Human Services*

supplemental funds for refugee resettlement support, hurricane relief, and responses to other infectious diseases, such as Zika and Ebola.

The department's budget submission is organized by PPAs. Although budgeted programs cover most HHS operational activities, the HHS budget does not have a common appropriations structure across programs.[17] For example, HHS does not receive separate appropriations for procurement and R&D; rather, most programs report different cost categories and mission requirements in their budget justifications. One HHS official observed that the distinction between different "colors of money" is simply not relevant to most OPDIV missions, and Congress provides HHS programs with considerable latitude in how to obligate appropriated funds across activities, such as procurement and R&D.[18] A key exception is capital investments, including information technology (IT) modernization and physical infrastructure, which are disbursed through department-wide working capital revolving funds (and are budgeted separately from program activities).[19] Whether for program or department-wide expenses, HHS follows an obligation-disbursement approach, in which funds must be obligated before they can be disbursed.

Most of HHS's discretionary operating budget is funded through one-year appropriations. However, certain discretionary activities—notably, biodefense, medical stockpiling, and medical countermeasure research at the NIH and CDC—are planned and budgeted with multiyear and no-year appropriations, the latter of which do not expire until they are expended.[20]

Supplemental appropriations vary significantly by obligation period, ranging from one-year to no-year funds depending on congressional intent and whether the funds have a designated purpose.[21] For example, the Coronavirus Preparedness and Response Supplemental Appropriations Act of 2020 included a mix of no-year (e.g., FDA salaries and expenses for activities "to prevent, prepare for, and respond to coronavirus" that were intended to "remain available until expended"), two-year, and four-year funds.[22]

HHS follows a budgeting framework that is common to most U.S. federal civilian agencies, beginning with an annual planning cycle and culminating in budget execution and performance evaluation. HHS budgeting policies and procedures are organized around OMB

(HHS), 2022: In Brief, Congressional Research Service, R47232, October 7, 2022, p. 1.

[17] HHS officials, interviews with the authors, October 2022–January 2023.

[18] HHS official, interview with the authors, November 2022.

[19] Some OPDIVs have their own working capital funds, but they can still apply for additional capital investment funds from the centralized accounts.

[20] HHS official, interview with the authors, October 2022.

[21] HHS official, interview with the authors, January 2023.

[22] Public Law 116-123, Coronavirus Preparedness and Response Supplemental Appropriations Act, 2020, March 6, 2020.

Circular No. A-11.[23] Although each OPDIV employs distinct approaches to budget planning and formulation, the standardization of policies and procedures at the headquarters level ensures some measure of stability and predictability from year to year and across administrations. The HHS budgeting process includes several key annual outputs with the approximate time frames shown in Table 3.2.

Preparing the HHS budget—from planning through budget presentation—generally takes between 18 and 24 months for any given fiscal year, as shown in Figure 3.2. However, in each calendar year, multiple cycles of planning, formulation, presentation, and execution lasting 18 months or longer are underway at any given time, as illustrated in the figure. (The approximate time frames in Table 3.2 roughly correspond to the first 12 months of the bottom bar in Figure 3.2.)

TABLE 3.2
HHS Budgeting Process

Key Annual Output	Approximate Time Frame
ASFR's Office of Budget (ASFR/OB) establishes OMB planning assumptions and sets fiscal guidance and funding targets. The HHS Secretary and OPDIV planning and strategy offices develop top-down and bottom-up priorities, respectively, for the upcoming budget.	February–March, Year $N - 2$
ASFR/OB sends guidance to the OPDIVs about planning assumptions and fiscal targets.	Late April–early May, Year $N - 2$
OPDIVs submit budget justifications to ASFR/OB with two or three different funding levels.	Late May–early June, Year $N - 2$
OPDIVs present budget justifications to the Secretary's Budget Council (SBC).	Mid-June–early August, Year $N - 2$
ASFR/OB incorporates feedback throughout the summer to develop a draft budget (i.e., straw budget), balancing OPDIV priorities with fiscal guidance.	July–August, Year $N - 2$
ASFR/OB reproposes a draft budget to the SBC, receives feedback, makes adjustments, and presents a draft budget to the HHS Secretary.	July–August, Year $N - 2$
HHS submits a draft budget to OMB. Operating and staff divisions hold budget hearings with OMB.	August–September, Year $N - 2$
OMB passes back recommendations on the draft budget.	November, Year $N - 2$
OPDIV financial management offices and CFOs develop congressional justification materials for congressional budget hearings.	January, Year $N - 1$
HHS uses a performance review process to self-assess implementation progress and the extent to which it has achieved its agency priority goals.	Quarterly

SOURCES: Features information from HHS officials, interviews with the authors, October 2022–January 2023.

---

[23] OMB Circular No. A-11, 2022.

FIGURE 3.2

Flowchart of the HHS Budgeting Process

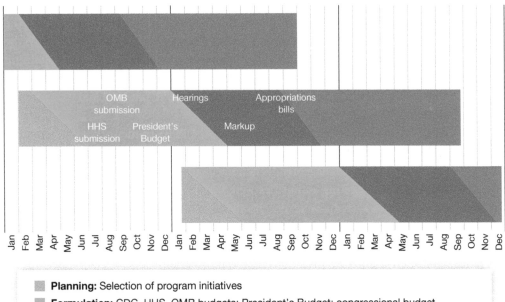

Planning: Selection of program initiatives

Formulation: CDC, HHS, OMB budgets; President's Budget; congressional budget

Presentation: Committee hearings, Q&As, House and Senate reports, appropriations bill

Execution: Apportionments, spend plans, allotments, allowances, and analysis and evaluation to inform the future strategic direction

SOURCE: Adapted from CDC, Financial Management Office, "Financial Management Course," presentation, undated.

In response to the guidance from President Johnson for federal civilian agencies to use program budgeting,[24] HEW created its PPBS in 1967. To institute the new process, which borrowed the core framework and organization from DoD, HEW created ASPE. HEW modeled ASPE on DoD's Office of Systems Analysis,[25] and HEW staffed ASPE with an analytic workforce that was responsible for program budgeting, strategic planning, performance evaluation, analysis of alternatives, and assisting the budget office in constructing five-year funding plans.[26] Additionally, HEW built an information system that linked planning information to mission priorities and program budgets.[27]

---

[24] Carl S. Rappaport, "Program Budgeting and PPBS in the Federal Government," paper presented at the 48th Annual Meeting of the Committee on Program Budgeting, 1969, p. 7.

[25] The Office of Systems Analysis evolved into the PA&E and, following the Weapon Systems Acquisition Reform Act of 2009, ultimately became CAPE in OSD.

[26] West, 2011, p. 29.

[27] Rivlin, 1969, p. 911.

In 1971, OMB suspended its requirement for all federal agencies to submit PPBS-level data, which effectively terminated the use of PPBS in most civilian agencies.[28] As of 1973, HEW was still following five-year budget plans and used ASPE to tie budgets to plans and departmental goals.[29] However, successive HEW secretaries in the late 1970s began unraveling the department's PPBS infrastructure, beginning with long-range planning and, ultimately, the link between plans and budgets. HEW gradually abandoned its strategic and budget planning processes during the Jimmy Carter administration and, thereafter, lacked a strategic planning capability until the passage and implementation of the GPRA in 1993 and its successor, the GPRA Modernization Act, in 2010.

At least three factors led HEW to move away from the PPBS framework. First, although the secretary had adopted program-based budgeting in the 1960s and centralized many planning- and budget-related management functions by the 1970s, HEW remained a highly federated organization. Budgeting authorities were fragmented across the department, and planning and budgeting systems were decentralized across subdepartment agencies and across programs.[30] The heavy emphasis on department funding for grants to state and local governments further decentralized budget transparency and accountability. Moreover, HEW programs were often funded by three different appropriations bills, which made linking program budgets to appropriations extremely complex and created challenges to adopting a common appropriations structure at the activities level (e.g., the colors of money, such as RDT&E).[31]

Second, key differences in DoD and HEW missions and activities led HEW officials to believe that PPBS was better suited to DoD.[32] ASPE officials argued that health services should not be treated like defense acquisition programs and that it was difficult for HEW to make the same budget trade-offs as DoD did.[33] Because the majority of HEW's programs

---

[28] Allen Schick, "A Death in the Bureaucracy: The Demise of Federal PPB," *Public Administration Review*, Vol. 33, No. 2, March–April 1973, p. 147.

[29] Schick, 1973, p. 156.

[30] See U.S. General Accounting Office, 1990, p. 22; and Beryl A. Radin, *Managing Decentralized Departments: The Case of the U.S. Department of Health and Human Services*, PricewaterhouseCoopers Endowment for the Business of Government, October 1999, p. 7.

[31] Rivlin, 1969, p. 913. According to the Congressional Research Service,

> [m]ost of the department's discretionary appropriations are provided through the Departments of Labor, Health and Human Services, and Education, and Related Agencies (LHHS) Appropriations Act. However, funding for certain HHS agencies and activities is provided in two other bills—the Departments of the Interior, Environment, and Related Agencies Appropriations Act and the Agriculture, Rural Development, Food and Drug Administration, and Related Agencies Appropriations Act. (Tollestrup, Lynch, and Cornell, 2022, pp. 4–5)

[32] Harlow, 1973, p. 90; Jablonsky and Dirsmith, 1978, p. 216.

[33] According to former HEW Assistant Secretary for Planning and Evaluation Alice Rivlin,

> [d]efense decisions result in some people being better protected or bearing a heavier burden than others, but these differential effects are not nearly so obvious as in domestic programs. In domestic programs of direct service to particular types of people, everyone knows who the immediate beneficiaries are.

employed a funding structure based on services and grants, ASPE eventually adopted the perspective that long-range budget planning was inappropriate.[34] As of the late 1960s, only a few R&D programs attempted to estimate future costs, which meant that the five-year funding plans were largely irrelevant to most HEW programs.[35]

Third, although ASPE was responsible for all planning, evaluation, and program-level analyses of alternatives in the department, the office lacked the staff and analytic capacity to handle such a broad mandate.[36] At the Secretary's direction in the late 1970s, ASPE shifted its focus from cost analysis, strategic planning, and evaluating budget trade-offs to health policy analysis and program evaluation.[37]

Given HHS's decision to move away from the PPBE construct in the 1970s, the contemporary HHS budgeting process differs in several important respects from that in such departments as DoD that employ a formal PPBE process. Although the HHS budgeting process begins with a planning component, budget planning is neither codified in agency directives nor institutionalized department-wide. Budget planning is aligned with the annual budget cycle and consists of an informal articulation of program and mission priorities at both the Secretary and OPDIV levels. HHS prepares a quadrennial strategic plan, but several HHS officials observed that this plan and the department's annual budget are developed independently and have few touchpoints, aside from being guided by similar leadership priorities.[38] Also unlike DoD, which prepares five-year funding plans, HHS generally does not conduct multiyear budget planning, with the exceptions previously noted. Because HHS organizes most of its programs around the annual budgeting process, there is an implicit disconnect between the department's long-range strategic planning and its program budgeting. Indeed, some HHS officials indicated in our interviews that a shift to a multiyear budgeting process that was more directly tied to strategic planning efforts would better inform resource allocation and could lead to improved program performance.[39]

The department's federated authorities and organization, dispersed appropriations sources, and disproportionate program orientation toward grants and services all help explain why HHS lacks a common appropriations structure comparable to DoD's. The service-oriented nature of HHS's funding likewise accounts for the department's reliance on

---

A good [PPBS] can illuminate these distributional decisions, but cannot make them any easier. Indeed, assembling and publicizing information on who is helped by particular government programs may intensify political conflict. (Rivlin, 1969, p. 922)

[34] U.S. General Accounting Office, 1990, pp. 22–25.

[35] Rivlin, 1969, p. 919.

[36] Rivlin, 1969, p. 917.

[37] U.S. General Accounting Office, 1990, pp. 22–25.

[38] HHS officials, interviews with the authors, October 2022–January 2023.

[39] HHS officials, interviews with the authors, October 2022–January 2023.

one-year appropriations (for discretionary spending) and its historical shift away from long-term budget planning.

Decisionmakers and Stakeholders

The HHS budgeting process combines bottom-up decisionmaking with top-down oversight. One HHS official described the process as "more bottom-up than top-down" at the program and activity levels: The HHS Secretary's leadership guides budget priorities and adjudicates among OPDIV submissions.[40]

According to the ASFR functional statement, that office exercises the "full Department-wide authority of the Secretary" in the areas of financial management and budget, as delegated by the HHS Secretary and approved in the Federal Register.[41] ASFR therefore functions as the HHS CFO, owns the annual budgeting process, and is the principal adviser to the HHS Secretary on the budget. The office is staffed by the OB, which provides initial fiscal and planning guidance to the OPDIVs at the start of budget formulation, reviews and evaluates OPDIV budget submissions, assists OPDIVs in completing their congressional justification materials, and advises the SBC. Meanwhile, ASPE owns the strategic planning process, serves as an advocate for agency strategic priorities on the SBC, and provides inputs for assessing progress toward HHS agency priority goals.[42]

Once the OPDIVs have submitted draft budget formulations to the OB, the SBC serves as a forum for OPDIV directors to present and defend their program priorities.[43] The SBC likewise enables the HHS Secretary's staff to assess the extent to which OPDIV submissions adhere to the agency priority goals, the Secretary's core strategic priorities and missions, and initial fiscal guidance.[44] The SBC is co-chaired by the HHS Deputy Secretary and the Assistant Secretary for Financial Resources, and it typically includes ASPE and the Office of the Assistant Secretary for Legislation (ASL).[45] Although the SBC is an informal decisionmaking body, it has become institutionalized since its creation in the early 1990s as an essential component in the budgeting process.[46] And because the HHS budgeting process is primarily bottom-up, the SBC serves as an important intermediary between the OPDIVs and the HHS Secretary. The SBC further constitutes a powerful source of authority: Both SBC co-chairs function as direct reports to the Secretary, conferring authority to the SBC to represent the

---

[40] HHS official, interview with the authors, October 2022.

[41] HHS, "Office of the Assistant Secretary for Financial Resources Functional Statement," webpage, last reviewed May 2, 2017, sections AM.00–AM.20.

[42] HHS, 2017; HHS officials, interviews with the authors, October 2022–January 2023.

[43] U.S. General Accounting Office, *Managing for Results: Efforts to Strengthen the Link Between Resources and Results at the Administration for Children and Families*, GAO-03-9, December 10, 2002, p. 16.

[44] HHS official, interview with the authors, November 2022.

[45] In some administrations, the HHS general counsel and chief of staff have also sat on the SBC.

[46] HHS officials, interviews with the authors, October 2022–January 2023.

Secretary's priorities. However, the department's organizational culture has also reportedly positioned the SBC as a collaborative, rather than adversarial, forum for the HHS Secretary to adjudicate among OPDIV budget priorities.[47]

The decentralized nature of HHS program budgets and appropriations creates multiple opportunities for HHS to interact with Congress. While the HHS Secretary has a direct relationship with congressional authorization and appropriations committees through both the OB and the ASL, each OPDIV presents its congressional justification on the Hill during budget hearings and maintains its own relationship with Congress. This arrangement creates the potential for a "multiple principals" problem in which the HHS Secretary retains ultimate authority over the department's budget submission but the OPDIVs have the opportunity to advocate to Congress on behalf of their own programs. Nevertheless, HHS officials described Congress as being sensitive to the Secretary's priorities and willing to provide the Secretary with significant budget flexibility to meet emerging public health needs.[48]

Planning and Programming

The unpredictability of public health crises, the service orientation of most HHS programs, and the annual HHS budget planning cycle significantly impede the department's ability to conduct long-term budget planning. The annual planning cycle is therefore ad hoc and primarily consists of an informal review of the prior-year budget, with input from the OB performance evaluation team.

Department-wide strategic priorities emanate from the quadrennial strategic planning process, which is led by ASPE (although many OPDIVs develop their own internal strategic plans). According to one HHS official, ASPE takes a "whole-of-department approach" to developing the strategic plan.[49] At the start of each new administration, HHS leadership creates a goal-to-objective framework and consults with department leadership to establish priorities. The department's draft goals and objectives are due to OMB in June, in accordance with OMB Circular A-11.[50]

Next, ASPE assigns strategic planning liaisons from every operating and staff division. ASPE conducts environmental scans with the division liaisons to assess changes to their missions and the effects of major legislation. Then, the office works with division subject-matter experts to compile the full narrative for the draft strategic plan. In recent years, this process has included more than 500 people across 21 working groups.[51] Following an internal ASPE

---

[47] During the Clinton administration, the HHS Secretary made the OPDIV directors sitting members of the SBC. In this role, the OPDIVs were able to comment on one another's budget proposals and program priorities (HHS official, interview with the authors, November 2022).

[48] HHS officials, interviews with the authors, October 2022–January 2023.

[49] HHS official, interview with the authors, November 2022.

[50] OMB Circular No. A-11, 2022.

[51] HHS official, interview with the authors, November 2022.

and department-wide review of the draft strategic plan, HHS submits the plan to OMB for consultation and public comment in September. ASPE works with the operating and staff divisions to incorporate public comments into the draft in the fall and then submits the final draft to OMB by the end of December.

Although the HHS strategic plan is the foundational articulation of department-wide goals and priorities, it only tangentially informs the budgeting process. At the start of each annual budget cycle, the HHS Secretary and OPDIV leadership identify key program priorities through separate (informal) planning processes.[52] At the department level, this planning process includes gathering inputs from the Secretary and ASPE, which may indirectly tie initial budget priorities to the strategic plan and agency priority goals. However, this link is weak; most HHS officials we spoke with indicated that the Secretary's budget priorities overwhelmingly emerge from a review of prior-year budgets as part of both ASFR/OB's ongoing performance evaluation activities and its periodic interactions with the OPDIVs' budget teams. ASPE's seat on the SBC creates another touchpoint for the HHS strategic plan in the budgeting process, but ASPE's role at this stage is primarily to advise the Deputy Secretary and ASFR in adjudicating strategic priorities among budget submissions rather than compelling adherence to budget priorities.

Each OPDIV has a unique budget planning process. For example, CDC financial resource planning occurs over three stages: advanced pre-planning, planning, and final and post-planning. In the advanced pre-planning stage, the CDC's Office of Policy and Strategy identifies known priorities and considers possible risks, incorporates forecasts of acquisition decisions, and assesses staffing requirements.[53] In the planning stage, CDC receives planning assumptions and fiscal guidance from ASFR/OB and develops spend plans. In the final and post-planning stage, CDC finalizes budget plans and transfers them into the Integrated Resources Information System (IRIS), the agency's consolidated budgeting and execution information system.

The two most important mechanisms for flowing information about operating and staff division budgetary needs to HHS headquarters are, arguably, the SBC consultations and the ASFR/OB performance evaluation process. During its consultations, the SBC allows the divisions to add context to their initial budget formulations and advocate for core agency missions and program needs. Aside from the more regular interactions between ASFR/OB and each division's budget office, the SBC serves as the primary conduit between division- and department-level leadership. For its part, the ASFR/OB coordinates with the OPDIVs to establish performance plans for each program in an enacted budget.[54] These performance targets serve as the basis for evaluating prior-year program performance in the subsequent budgeting cycle and are an important feedback mechanism.

---

[52] HHS officials, interviews with the authors, October 2022–January 2023.

[53] HHS officials, interviews with the authors, October 2022–January 2023.

[54] HHS official, interview with the authors, October 2022.

Budgeting and Execution

The ASFR functional statement delegates authority for the budgeting process to ASFR.[55] ASFR/OB oversees the budget formulation phase and is responsible for issuing guidance and setting funding targets, developing planning assumptions, reviewing budget submissions, and serving as the HHS Secretary's principal liaison to the operating and staff divisions in the budget formulation phase.[56] In accordance with OMB Circular No. A-11, the HHS budgeting process is concerned primarily with developing a one-year budget submission, including justifications and presentation materials, that results in a congressional appropriation.[57]

The budgeting formulation phase begins after the HHS Secretary and OPDIVs have separately identified key priorities and planning assumptions for the upcoming budget. Using those key priorities and planning assumptions, each operating and staff division prepares a budget plan that serves as the basis for its budget submission.[58] By early June, each HHS division submits a unique budget justification to the OB for review. ASFR/OB holds an internal budget review, with at least one analyst assigned to each budget account.[59] As part of this initial budget review, ASFR/OB considers the fiscal implications of each proposed justification, assesses the rationale for any significant new proposals, and considers whether the proposed justification is likely to gain traction with Congress.[60]

Following the internal budget review, each OPDIV (usually its agency director) presents its budget justification to the SBC. (While some staff divisions also present their budget justifications to the SBC, others are handled by ASFR/OB.)[61] The SBC serves as a forum for OPDIVs to both convey information on key fixed costs (e.g., pay, continuations of grants) and share opportunities that can have the greatest impacts on their missions.[62] The SBC can ask for clarification and context for how the OPDIV budget plans are intended to address the Secretary's key priorities and HHS strategic goals and missions.

ASFR/OB staffs the SBC and schedules follow-up meetings with the OPDIVs to incorporate SBC feedback into the budget justifications. Depending on the outcomes of the budget presentations and continuous feedback from the OPDIVs throughout summer, ASFR/OB develops a draft HHS budget (called the *straw budget*). The straw budget accounts for pro-

---

[55] HHS, 2017, sections AM.00–AM.20.

[56] HHS official, interview with the authors, October 2022.

[57] OMB, 2022.

[58] HHS official, interview with the authors, October 2022.

[59] HHS official, interview with the authors, October 2022.

[60] HHS official, interview with the authors, November 2022.

[61] HHS official, interview with the authors, November 2022.

[62] HHS official, interview with the authors, November 2022.

posed changes to prior-year program budgets, subject to OMB fiscal targets.[63] ASFR/OB submits the draft budget to the SBC and goes through one additional round of SBC-driven revisions. Then, ASFR/OB presents the draft budget to the HHS Secretary, incorporating the Secretary's feedback as necessary. Finally, by September, HHS submits its draft budget to OMB for review, operating and staff division budget hearings are scheduled, and OMB completes budget passback with HHS.[64]

As discussed, discretionary HHS funds are overwhelmingly budgeted annually, but some discretionary programs receive multiyear or no-year appropriations. Congress appropriates discretionary multiyear funding to some biodefense, medical stockpiling, strategic preparedness, and medical countermeasure development programs,[65] but these appropriations are limited to only a few R&D programs in NIH, CDC, and the FDA.[66] The other major sources of multiyear and no-year discretionary funds are supplemental appropriations, which have different obligation periods depending on congressional intent and whether Congress has identified a dedicated purpose for the funds. Supplemental funding for hurricane relief, for instance, is typically dedicated to specific uses within a limited obligation window, whereas supplemental appropriations for public health emergencies, such as COVID-19, often give HHS significant flexibility and no-year funding.[67]

Some OPDIVs (particularly the FDA) also employ fee authorities. These fees may offset appropriations and are justified accordingly in budget submission materials.

Mandatory programs are funded through a mix of authorizing statutes and annual appropriations (in the case of Medicaid, for example). Although the annual HHS budgeting process does not alter funding for programs appropriated by statutes and funding processes outside the annual congressional appropriations process, the HHS budget justifications do include narratives highlighting material changes in the status of program costs and benefits.

As in the budget formulation phase, the execution phase features a mix of top-down and bottom-up decisionmaking authorities. Once funds are appropriated, ASFR/OB's Division of Budget Policy, Execution, and Review manages the apportionment process from the top down.[68] This division's core responsibilities include providing "direction in the Departmentwide review, analysis and appraisal of financial elements of program execution and the development and execution of policies related to efficient allocation, expenditure and control of

---

[63] Fiscal targets are more concrete for discretionary programs. Fiscal targets for mandatory programs tend to be less defined because OMB does not always provide mandatory budget targets. Therefore, the mandatory targets are often more thematic, such as savings proposals in Medicare focused on adding competition or improving Medicare benefits and operations (HHS official, interview with the authors, November 2022).

[64] *OMB passback* refers to a prioritization process between OMB and HHS in which OMB's focus is on ensuring that the President's budgetary priorities are addressed.

[65] HHS officials, interviews with the authors, October 2022–January 2023.

[66] HHS officials, interviews with the authors, October–November 2022.

[67] HHS official, interview with the authors, January 2023.

[68] HHS official, interview with the authors, January 2023.

funds."[69] ASFR/OB's Division of Budget Policy, Execution, and Review "[c]oordinates and tracks outlay projections" and monitors the Unified Financial Management System (UFMS), an information system maintained by the Program Support Center that tracks department-wide budget execution.[70]

Simultaneous with this top-down budget execution management, however, each operating and staff division manages the obligation of its own apportioned funds and employs unique execution monitoring systems. This situation has led to five different sets of books in HHS, which could complicate the department's budgeting process or at least render it inefficient.[71] Still, ASFR/OB retains some top-down oversight and control through its management of the apportionment process, its ability to monitor outlays through the UFMS, and its requirement that the operating and staff divisions submit monthly reports on apportionments and allowances all the way down to invoices.[72]

During the execution phase, programs may require additional discretionary funding to cover emerging needs or unforeseen crises. The LHHS appropriations bills afford HHS considerable flexibility in reprogramming funds: Below a reprogramming threshold of the lesser of $1 million or 10 percent of a budget account, HHS is not required to report to Congress.[73] Above that threshold, HHS must notify Congress of any reprogramming; additionally, it must notify Congress above a threshold of $500,000 if, for example, the reprogramming decreases a program's appropriated funding by more than 10 percent or substantially affects the program's personnel or operations. Congress has provided HHS with additional sources of flexibility during budget execution, including the Secretary's One-Percent Transfer General Provision, which allows HHS to transfer up to 1 percent from any LHHS appropriation account into another account up to 3 percent of the amount of that receiving account, with a maximum transfer amount of around $900 million.[74]

Other flexible-spending accounts include the No-Surprises Implementation Fund, which is distributed across multiple departments; the COVID-19 Emergency Fund; CDC's Infectious Disease Rapid Response Reserve Fund; the Public Health and Emergencies Account for emergency supplemental appropriations; and the Non-Recurring Expenses Fund (NEF). Congress proposed the NEF in 2008 as a mechanism to obligate appropriated funds more efficiently and to address critical department-wide technology and infrastructure needs.

---

[69] HHS, 2017, section AML4.

[70] The Program Support Center is a shared services organization within ASA that provides program and administrative support to the entire department. HHS official, interview with the authors, January 2023; HHS, 2017, section AML4.

[71] HHS official, interview with the authors, January 2023.

[72] HHS official, interview with the authors, January 2023.

[73] NIH, Office of Management and Assessment, "Budget Execution," Chapter 1920, *NIH Policy Manual*, March 10, 2020.

[74] The Office of Refugee Resettlement can take up to 15 percent of the value of transfer funds, so these funds are often transferred there (HHS official, interview with the authors, January 2023).

Multiple HHS officials identified the NEF as an authority that provides vital budgetary flexibility to the department.[75] Under the NEF, HHS can take expired, unobligated funds and reallocate them to a department-wide capital investment account. Importantly, the funds may not be reobligated to their original purpose. HHS has used the NEF extensively to fund IT systems, particularly for cybersecurity purposes, but operating and staff divisions can request funding for capital expenditures.

If current-year appropriations are passed under a continuing resolution, HHS operations would be suspended for new contracts and grants across all discretionary programs. In practice, this means that HHS must work within the prior fiscal year account structure. Continuing resolutions tend to be most damaging for HHS's R&D programs that operate on multiyear funds because program managers are unable to complete contract negotiations on ongoing R&D process work. Continuing resolutions also disproportionately harm divisions that incur high salary and operating costs, as well as those that rely on contractors to fulfill their missions and those with high degrees of staffing uncertainty.[76] To mitigate the impact of a continuing resolution on agency operations, and in accordance with Section 124 of OMB Circular No. A-11, HHS prepares an annual contingency staffing plan.[77] The number of full-time HHS staff expected to be retained during a lapse of appropriations varies from year to year, ranging from 60 percent in FY 2023 to 76 percent in FY 2019.[78] However, OPDIVs "with a substantial direct service component," such as the NIH Clinical Center, tend to retain more staff during a continuing resolution.[79] In general, the impact of continuing resolutions is limited department-wide because most HHS programs operate on one-year funds. Furthermore, HHS officials indicated that grantmaking entities plan for continuing resolutions by developing requests for proposals in advance, which lessens the impact unless the continuing resolution extends deeper into the fiscal year.[80]

Performance and program evaluation in HHS are conducted through multiple channels, each of which creates feedback mechanisms at multiple points in the budgeting process. ASFR/OB's Division of Budget Policy, Execution, and Review leads performance assessment activities, including the development of performance measures.[81] Its performance evaluation role feeds directly into the budgeting process during ASFR/OB's internal reviews of the divisions' budget submissions. For example, ASFR/OB works with the operating and staff divisions during budget execution to establish performance plans and targets to inform the sub-

---

[75] HHS officials, interviews with the authors, October 2022–January 2023.

[76] HHS official, interview with the authors, November 2022.

[77] OMB, 2022.

[78] HHS, "FY 2023 HHS Contingency Staffing Plan for Operations in the Absence of Enacted Annual Appropriations," webpage, last reviewed October 5, 2022b.

[79] HHS, 2022b.

[80] HHS officials, interviews with the authors, October 2022–January 2023.

[81] HHS, 2017, section AML4.

sequent budget cycle. The Division of Budget Policy, Execution, and Review also includes the HHS performance improvement officer, who is responsible for coordinating delivery of the department's annual performance plan to Congress in accordance with GPRA and the GPRA Modernization Act of 2010.[82] For its part, ASPE serves as the HHS evaluation officer and is responsible for fulfilling the statutory requirements in the Foundations for Evidence-Based Policymaking Act of 2018 (also known as the Evidence Act).[83] ASPE provides inputs from program evaluation activities to the budgeting process through its strategic planning role but also as a member of the SBC. In general, however, ASPE's inputs to the budgeting process are more indirect relative to the other inputs.

Oversight

HHS's budgeting process takes between 18 and 24 months from the beginning of the planning phase to the submission of the budget request, congressional justifications, and execution. Along the way, the multiple reviews by HHS leadership and OMB provide opportunities to adapt the budget request to meet emerging public health threats and evolving missions.

Once the President's Budget is released (which by law should occur in early February), the House and Senate authorization and appropriations committees begin reviewing it. Because most of HHS's discretionary programs and activities are separately authorized under the Public Health Service Act of 1944, HHS typically does not receive annual authorizing legislation for its appropriations. Therefore, review of the President's Budget for HHS is primarily performed by the House and Senate appropriations committees and subcommittees on labor, health and human services, and education.[84] For HHS, this review occurs through congressional staff briefings with HHS and OPDIV officials and culminates in committee hearings. The House and Senate committees conduct budget hearings with the HHS Secretary each year. The appropriations committees hold additional hearings with the operating and staff division heads, typically in March and April. The subcommittees then typically release the text of the three HHS appropriations bills in the summer and add subcommittee and full committee markups. For full committee markups, the committees attach accompanying reports to the bill, which can provide additional guidance. In recent years, HHS appropriations have been routinely passed via continuing resolutions or omnibus (i.e., consolidated) appropriations bills.

---

[82] Pub. L. 103-62, 1993; Public Law 111-352, GPRA Modernization Act of 2010, January 4, 2011.

[83] Pub. L. 115-435, 2019; HHS officials, interviews with the authors, November 2022.

[84] HHS receives funding through three separate appropriations bills: primarily through the LHHS Appropriations Act and secondarily through the Departments of the Interior, Environment, and Related Agencies Appropriations Act and the Agriculture, Rural Development, Food and Drug Administration, and Related Agencies Appropriations Act. The House and Senate appropriations subcommittees on labor, health and human services, and education are responsible for the majority of discretionary funds that are appropriated to HHS.

HHS maintains close oversight of funds execution through the UFMS. As described earlier, ASFR/OB manages the allocation of appropriated funds. Although oversight of the obligations is decentralized to the OPDIVs, ASFR/OB monitors obligation rates through the UFMS and monthly execution reports prepared by the OPDIVs. Therefore, HHS retains relatively centralized oversight of the execution phase despite the distribution of authorities across the department. ASFR's Office of Finance also conducts an annual audit, which involves sending out automated questionnaires two or three times per year.[85]

Analysis of HHS's Budgeting Process

Strengths

We identified six strengths in HHS's budget process. First, the department is afforded several *flexibility mechanisms* in budgeting. These mechanisms are responses to both the unpredictability of HHS's missions and the need for additional sources of funding for IT modernization and capital investments outside HHS's one-year program funds.[86] To this end, Congress has created several flexible-spending accounts, such as the CDC's Infectious Disease Rapid Response Reserve Fund and the Public Health Service Evaluation Set-Aside, to enable HHS to adapt quickly to emerging public health threats. Meanwhile, the NEF allows HHS to reallocate expiring, unobligated funds to capital investments—something that HHS has used to bolster its investments in IT modernization and cybersecurity—outside the traditional appropriations process.

Second, Congress historically provides significant funds through *supplemental appropriations* to address public health emergencies. Congress has passed seven COVID-19 supplemental appropriations bills since 2020, along with other supplemental bills for hurricane relief and Ebola response. Moreover, supplemental appropriations are often accompanied by significant spending flexibility through multiyear or no-year implementation funds.

The third strength has been *consistency* in its policies and processes. Roles and responsibilities are clearly articulated in functional statements approved by the ASA, and resource management follows a predictable annual cycle. Budget planning, formulation, and execution processes have remained stable since the late 1970s. Although each OPDIV employs distinct approaches to budget planning and formulation, the standardization of policies and procedures in accordance with OMB Circular No. A-11 at the headquarters level ensures some measure of stability and predictability from year to year. Additionally, HHS prepares an annual contingency staffing plan to minimize the impact of congressional continuing resolutions on agency operations.

---

[85] HHS official, interview with the authors, January 2023.

[86] For example, HHS has used the NEF for facilities infrastructure acquisitions, such as a new wing at the NIH facility in Rockville, Maryland, to provide new research laboratory space (HHS, "Fiscal Year 2017 Nonrecurring Expenses Fund: Justification of Estimates for Appropriations Committee," undated).

Fourth, the SBC is viewed internally as an extremely *effective forum* for adjudicating budget priorities and trade-offs. Budget presentation to the SBC is a deliberative yet two-way decisionmaking process, guided by a top-down structure that nevertheless allows OPDIVs and programs to advocate for their missions and fixed costs that must be met. Accountability over the budget process is centralized to the HHS Deputy Secretary, who co-chairs the SBC. Thus, the SBC has been institutionalized as a collaborative, rather than adversarial, forum for adjudicating budget decisions.

Fifth, HHS is effective at *managing continuing resolutions*. Given the frequency with which HHS has operated under a continuing resolution since 2000, most divisions and programs have developed contingency plans. HHS also prepares an annual department-wide contingency plan. Grantmaking agencies all do advance work (e.g., draft requests for proposals) to prepare for continuing resolutions and are able to function more or less normally during and after them. HHS's relatively stable levels of discretionary spending also help it avoid most of the contracting and acquisition challenges that occur in DoD, which often initiates new or innovative programs in each budget and is also subject to considerable budget instability and uncertainty. HHS's effective continuing resolution management enables it to retain anywhere from two-thirds to three-fourths of its full-time staff during lapses in appropriations.[87]

Finally, HHS has strong *transparency and oversight mechanisms* despite being a heavily decentralized organization. ASFR/OB manages budget execution department-wide and has relatively sweeping monitoring capabilities. It also conducts an annual audit through automated financial management systems, providing department-wide assurance throughout the decentralized parts of the execution process.

Challenges

We identified three challenges in HHS's budget process. First, annual budgets are largely *disconnected from long-term and strategic planning.* ASPE, as owner of the HHS strategic plan, provides planning and program evaluation inputs to the SBC. But, otherwise, ASPE has a limited role in overseeing the budget's construction. Annual budgets build on prior-year justifications and include emerging priorities, but budgeting at HHS lacks long-term vision.

Second, HHS *mostly uses one-year discretionary funds*, with about 80 percent of the discretionary obligations made in September.[88] Some HHS officials expressed a desire to operate on two-year discretionary funds to provide greater flexibility for contracting and execution, as well as for managing continuing resolutions. Only a few programs in the base discretionary budget are appropriated multiyear funds.[89] The short-term nature of HHS discretionary funding reflects the orientation of its programs toward health services and grants. But the

---

[87] HHS, 2022b.

[88] HHS official, interview with the authors, January 2023.

[89] HHS officials, interviews with the authors, October 2022–January 2023.

nature of this funding also places constraints on contracting and budget execution and limits the budget planning horizon of most HHS programs to a year-to-year basis.

Third, HHS's *federated organizational structure limits transparency.* HHS headquarters maintains some oversight of program execution through the UFMS and monthly execution reports, but there is neither department-wide control nor standardized spend-rate targets. HHS does not have a consolidated data system for budget formulation, which inhibits visibility into OPDIV budgets and budget planning. Each OPDIV uses a different set of financial books, which creates inefficiencies in department-wide financial management. Finally, it is difficult to incorporate performance evaluation into budgeting, given the high share of funding across OPDIV program activities that is devoted to grants and delegated to state, local, and tribal agencies.

Applicability

In the 1960s, HHS built an analytic workforce and organizational capacity that was modeled on DoD's PPBE system; however, in the 1970s, HHS gradually abandoned its own PPBE infrastructure. This shared legacy with PPBE highlights key differences between DoD and HHS. Because much of HHS's discretionary budget authority is for grants and services, HHS has increasingly relied on one-year funding for discretionary programs. This tendency has, in turn, led HHS leadership over time to disconnect long-term planning from the budgeting process. HHS and DoD therefore employ fundamentally distinct structures for funding their programs and formally linking their planning and budgeting processes.

HHS still uses program-based budgeting, but it has never adopted a common appropriations structure, both because HHS appropriations are derived from three different appropriations bills and because HHS programs and authorities are decentralized across the department. Moreover, the service-oriented nature of most HHS programs has precluded the creation of formal requirements and acquisition processes akin to DoD's Joint Capabilities Integration and Development System (JCIDS) or the DoD 5000 series of policy guidance. Key exceptions include R&D programs (especially at NIH and CDC), some of which use multiyear funds, and capital investments and IT modernization, which are often funded through the NEF and department-wide working capital revolving funds.

There are, however, some points of similarity between the DoD and HHS resource management systems. HHS's ASPE was originally modeled on DoD's Office of Systems Analysis, and the division retains much of its initial focus on program evaluation. Likewise, the HHS OB uses performance evaluation as an input of the budgeting process in a way that is somewhat comparable to that of CAPE in DoD's programming phase.

Lessons from HHS's Budgeting Process

Lesson 1: Collaborative Top-Down and Bottom-Up Budgeting Increases Alignment with Missions and Priorities

Because HHS does not use a common appropriations structure, budget justifications focus heavily on missions and needs. This focus allows discussions between the OPDIVs and the SBC's department-level leadership to concentrate on aligning program budgets and missions with the Secretary's priorities. DoD's PPBE System is often criticized for the adversarial atmosphere that emerges from the evaluation of program trade-offs conducted by OSD. While some competition helps produce effective military solutions, it can be overdone. In contrast, the SBC has created a collaborative forum for adjudicating budget trade-offs that is focused on executing organizational missions rather than competing for scarce resources.

DoD is already taking steps to address some aspects of this issue. For example, DoD initiated a pilot program in 2020 for a new RDT&E budget activity (BA-8) that is designed to centralize funding for software development across different colors of money.[90] HHS's experience with this sort of holistic program budgeting suggests that it is an important tool to enable agency leadership to prioritize missions and needs in the budget. DoD could also consider whether its program review and decision activities in the 3-Star Group and the Deputy Secretary–chaired Senior Leader Review Group, which address major issues during the programming phase, are sufficiently transparent and collaborative. As in HHS, DoD adjudicates budget trade-offs at a high level: The Chairman of the Joint Chiefs of Staff is responsible for preparing a threat analysis of DoD components' initial program objective memorandum submissions, and the director of CAPE conducts a program review and submits decision memoranda in collaboration with the DoD Comptroller.[91] This approach to budget adjudication is explicitly top-down and excludes DoD component leadership from deliberations. Consequently, DoD's budget adjudication process is less collaborative than HHS's, which has resulted in the perception of DoD's programming phase as a competition for scarce resources. By excluding bottom-up participation in program decisions, DoD may also obscure important details relating to components' missions, needs, and fixed costs from its annual budget submissions. Indeed, a primary rationale for HHS to include OPDIV directors in the SBC's budget deliberations is to add context to the budget and ensure that agency missions are adequately represented in forums for the services to present their mission needs and budget plans.

---

[90] Office of the Under Secretary of Defense for Acquisition and Sustainment, "Budget Activity (BA) 'BA-08': Software and Digital Technology Pilot Program Frequently Asked Questions," version 1.0, U.S. Department of Defense, September 28, 2020.

[91] DoDD 7045.14, 2017.

Lesson 2: Budget Flexibility Enables Responsive Funding Reallocation

DoD, like HHS, faces considerable unpredictability in the evolving nature of its missions and the threat environment. However, Congress has provided HHS with significant flexibility to manage and allocate appropriated funds. In recent years, HHS has received hundreds of billions of dollars in emergency supplemental funding to support its response to emergent public health threats. Moreover, HHS operates several flexible-spending accounts, such as the NEF, which allow HHS to reallocate expired, unobligated funds to capital investments. Although most of HHS's base discretionary funding consists of one-year appropriations, these flexibility mechanisms are often given as multiyear or no-year funding.

Our interviews with HHS officials suggest that much of HHS's budgetary flexibility stems from its relationship with Congress. HHS has tied its flexibility mechanisms to specific missions and purposes to ensure proper oversight and compliance with congressional intent. For example, the NEF does not allow reobligated funds to be used for their original purposes; rather, expired funds must be used for capital investments. Notably, DoD had a mechanism for reallocating expiring, unobligated funds that was similar to the NEF until the early 1990s.[92] HHS's experience indicates that Congress may, therefore, be willing to provide budget flexibility so long as it is tied to specific purposes designed to improve department-wide operations. Congress has also been particularly responsive to the uncertainty facing HHS missions and operations. Many of Congress's supplemental emergency appropriations to HHS build in considerable flexibility to allow HHS to address the multidimensional, evolving nature of public health crises. In contrast, DoD's supplemental funding for ongoing contingency operations has often been criticized as a slush fund. HHS has largely avoided this problem by maintaining a transparent, mutually supportive relationship with congressional appropriators and by emphasizing mission narratives and the unpredictability of HHS operations in its discussions with Congress.[93]

Lesson 3: Centralized Oversight Mechanisms Boost Accountability

Although HHS does not have a consolidated resource management information system, some OPDIVs have constructed such systems. For example, CDC uses IRIS to track its budgeting process from the planning phase through execution. At the department level, the lack of a consolidated budget formulation system has left HHS leadership with limited visibility into the OPDIVs' budgets. In contrast, ASFR/OB has fairly strong oversight of department-wide budget execution (or, in the case of program obligations, at least the ability to monitor OPDIV obligation rates). Finally, the OB's annual audit instills accountability over HHS's entire financial management enterprise.

---

[92] HHS official, interview with the authors, January 2023.

[93] HHS officials, interviews with the authors, October 2022–January 2023.

Because DoD is a comparably federated organization with diffuse authorities, it could explore the feasibility, costs, and benefits of constructing a consolidated PPBE information system. Like HHS, DoD lacks transparency at the headquarters level into component-level budget planning and formulation processes. However, although DoD implements department-wide standards for program obligation and execution rates, DoD components maintain unique execution and financial management systems that inhibit both a clean audit and real-time, department-wide monitoring of funds execution. A consolidated PPBE information system would arguably address most of these shortcomings, potentially leading to greater efficiencies and oversight in the DoD PPBE process.

Table 3.3 summarizes these three lessons.

TABLE 3.3

Summary of Lessons from HHS's Budgeting Process

Theme	Lesson Learned	Description
Budgeting and execution	Lesson 1: Collaborative top-down and bottom-up budgeting increases alignment with missions and priorities.	DoD could analyze the effectiveness of the 3-Star Group and Senior Leader Review Group as deliberative forums for adjudicating budget decisions that are collaborative and mission-focused. It could also expand on the new BA-8 pilot program to ensure more-holistic program budgeting.
Budgeting and execution	Lesson 2: Budget flexibility enables responsive funding reallocation.	Congress has provided HHS with several budget flexibility mechanisms that DoD could benefit from, particularly the One-Percent Transfer General Provision and the NEF. These mechanisms, along with HHS's supplemental appropriations, are a function of HHS's mission-focused budgets and its collaborative relationship with Congress.
Oversight	Lesson 3: Centralized oversight mechanisms boost accountability.	DoD could explore the feasibility, costs, and benefits of implementing a consolidated PPBE information system.

National Aeronautics and Space Administration

Sarah W. Denton and William Shelton

NASA was established in 1958 and is responsible for U.S. aeronautical and space activities.[1] As a civilian agency, NASA is devoted to peaceful activities that benefit humankind. It describes its mission as follows: "NASA explores the unknown in air and space, innovates for the benefit of humanity, and inspires the world through discovery."[2] The National Aeronautics and Space Act of 1958 outlined eight activities that support this mission:[3]

- Expand human knowledge of atmospheric and space phenomena.
- Improve the "usefulness, performance, speed, safety, and efficiency of aeronautical and space vehicles."
- Develop and operate "vehicles capable of carrying instruments, equipment, supplies, and living organisms through space."
- Establish long-range studies on the potential opportunities and challenges for use of "aeronautical and space activities for peaceful and scientific purposes."
- Preserve "the role of the United States as a leader in aeronautical and space science and technology and in the application to peaceful activities within and outside the atmosphere."
- Share information with national defense agencies on "discoveries that have military value or significance."
- Cooperate with other nations and groups of nations on peaceful aeronautical and space activities.
- Effectively use "scientific and engineering resources of the United States," in close cooperation with applicable federal agencies, toward peaceful ends.

---

[1] Pub. L. 85-568, 1958.

[2] NASA, 2022e.

[3] Pub. L. 85-568, 1958, p. 427.

These activities are supported by NASA's core values of safety, integrity, teamwork, excellence, and inclusion. According to NASA, these values are embedded in its self-ascribed attributes, shown in Figure 4.1.[4]

NASA has more than 18,000 civil servants from a multitude of disciplines with various backgrounds who work with many more contractors, academics, international colleagues, and commercial partners to accomplish its mission. Overall, NASA's workforce comprises more than 312,000 professionals at 20 centers and facilities across the United States.[5] Figure 4.2 shows NASA's organizational and reporting structure.

NASA's core values and attributes appear to influence its approaches to PPBE—especially its team-oriented processes. NASA's culture is highly collegial, which promotes engagement up and down the organization. The entire budgeting process—from development to execution and performance management—is overseen by the OCFO through representatives at NASA headquarters, the Mission Support Directorate, five mission directorates, and ten geographically dispersed centers, including the Jet Propulsion Laboratory, NASA's federally

FIGURE 4.1

NASA's Attributes

Curiosity

"Continually asking questions and seeking answers. Every day is a chance to try something new and come up with novel solutions to 'unsolvable' problems."

Team-oriented

"Come together as one to solve complex issues. Innovation is a staple, teamwork is a must and everyone's opinion counts."

Excellence

"Continuously striving to be better and know more. Pursue excellence in both the ordinary and the extraordinary."

Passion for exploration

"Constantly embarking on a range of adventures to better understand our planet, the solar system and beyond."

Agility

"Comfortable and flexible working in ambiguous environments. We embrace change and are ready to grow and adapt to what the future may bring."

Resilience

"Don't give up. We aren't deterred by obstacles or constraints, and we stay the course to achieve our goals. "

SOURCE: Features information from NASA, 2022e. Also see NASA Policy Directive (NPD) 1000.0C, *NASA Governance and Strategic Management Handbook*, Office of the Associate Administrator, National Aeronautics and Space Administration, January 29, 2020.

---

[4] NASA, 2022e.

[5] NASA, 2023a.

FIGURE 4.2
NASA's Organizational Structure

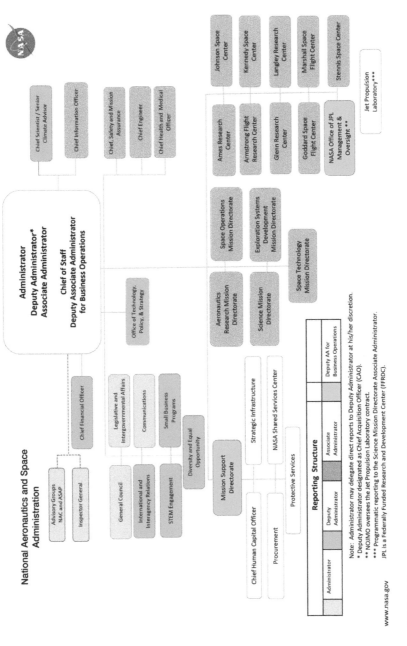

SOURCE: Reproduced from NASA, 2022d.

NOTES: AA = Associate Administrator; ASAP = Aerospace Safety Advisory Panel; JPL = Jet Propulsion Laboratory; NAC = NASA Advisory Council; NOJMO = Office of JPL Management and Oversight.

funded research and development center.[6] NASA's annual budget generally amounts to less than 0.5 percent of the entire U.S. federal budget, and it received $23.27 billion in funding in FY 2021. By comparison, NASA generated economic output of more than $71.2 billion in FY 2021, and this output resulted in approximately $7.7 billion in federal, state, and local tax revenues that year.[7]

NASA's organizational structure is comparable to that of DoD. For example, NASA headquarters acts similarly to OSD; the mission directorates resemble the military services; and NASA's centers function like other DoD agencies, such as the Defense Information Security Agency and Defense Logistics Agency.[8] Despite NASA's comparatively smaller annual budget, its approach to budgeting could provide useful insights for DoD about budgeting flexibility and agility.

NASA's budgeting process has a variety of attributes that we discuss in more detail in the remainder of this chapter:

- NASA's appropriations differ from DoD's.
- All appropriations are for two years, except for construction, which is funded on a six-year basis.
- Funding is budgeted primarily at the program level or above.
- Cost estimation and budget analysis functions are housed in NASA's Strategic Investment Division (SID) of OCFO.
- NASA's directorates are its budgeting centers of gravity.

Figure 4.3 shows the geographic distribution of NASA's headquarters, centers, and supporting facilities. Although NASA is headquartered in Washington, D.C., it has a presence across the United States.

Overview of NASA's Budgeting Process

SID produces NASA's strategic plan, which describes how the agency intends to "explore the secrets of the universe for the benefit of all."[9] NASA's 2022 strategic plan informed its FY 2023 budget request, which totaled $25.97 billion. Figure 4.4 shows how NASA's budget soared in the 1960s, fell in the 1970s, and has slowly rebounded ever since (in constant 2021 dollars).

---

[6] Note that NASA's PPBE process is outlined in a single publicly available document: NPR 9420.1A, 2021.

[7] NASA, 2022f.

[8] The mission directorates provide the centers funding for program-specific activities. Other center funding is provided via the Mission Support Directorate via two budget lines: safety, security, and mission services (SSMS) and construction and environmental compliance and restoration (CECR). The SSMS account covers work that is not specific to any given center's programs or projects (e.g., security officers). CECR covers all construction work related to center programs and projects.

[9] NASA, *FY 2023 President's Budget Request Summary*, 2022c; NASA, *2022 Strategic Plan*, 2022a.

FIGURE 4.3
NASA's Geographically Dispersed Centers and Facilities

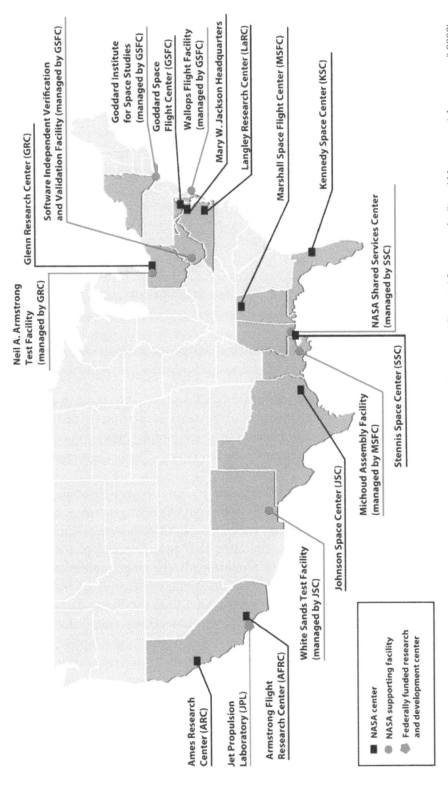

SOURCE: Reproduced from NASA, *FY 2022 Agency Financial Report*, Section 3, "Summary of Financial Statement Audit and Management Assurances," 2022b, p. 1.

FIGURE 4.4
NASA's Budget Authority over Time

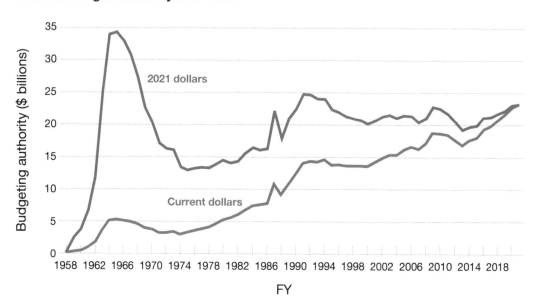

SOURCE: Adapted from Daniel Morgan, *NASA Appropriations and Authorizations: A Fact Sheet,* Congressional Research Service, July 2, 2021, Figure 1.

NASA's budgeting follows a four-phase PPBE process. The agency has organized its resource allocation guidance into two primary guidance documents, the first of which covers the first three phases: budget requirements development (planning), request for appropriations (programming), and budget formulation (budgeting),[10] while the second guidance document covers execution.

OCFO—specifically, through SID and the Budget Division—controls NASA's major budget formulation processes. Figure 4.5 outlines NASA's PPBE process.

During the planning phase, agency leaders establish NASA-wide policies and priorities, initial budget controls, and full-time equivalent staffing targets; update strategic plans; and provide top-level guidance for budget formulation.[11] The programming phase consists of analyzing programmatic priorities and needs, developing resourcing requirements, and identifying budgeting challenges.[12] The budget request is finalized during the budgeting phase. OCFO engages with OMB on the initial request and subsequent revision, which culminates in

---

[10] NPR 9420.1A, 2021. The final phase, execution—which includes financial management practices for budget authority, spending, recording, controlling, and reporting—is covered in a separate document: NPR 9470.1, *Budget Execution,* Office of the Chief Financial Officer, National Aeronautics and Space Administration, December 24, 2008.

[11] NPR 9420.1A, 2021.

[12] NPR 9420.1A, 2021.

FIGURE 4.5
NASA's PPBE Process

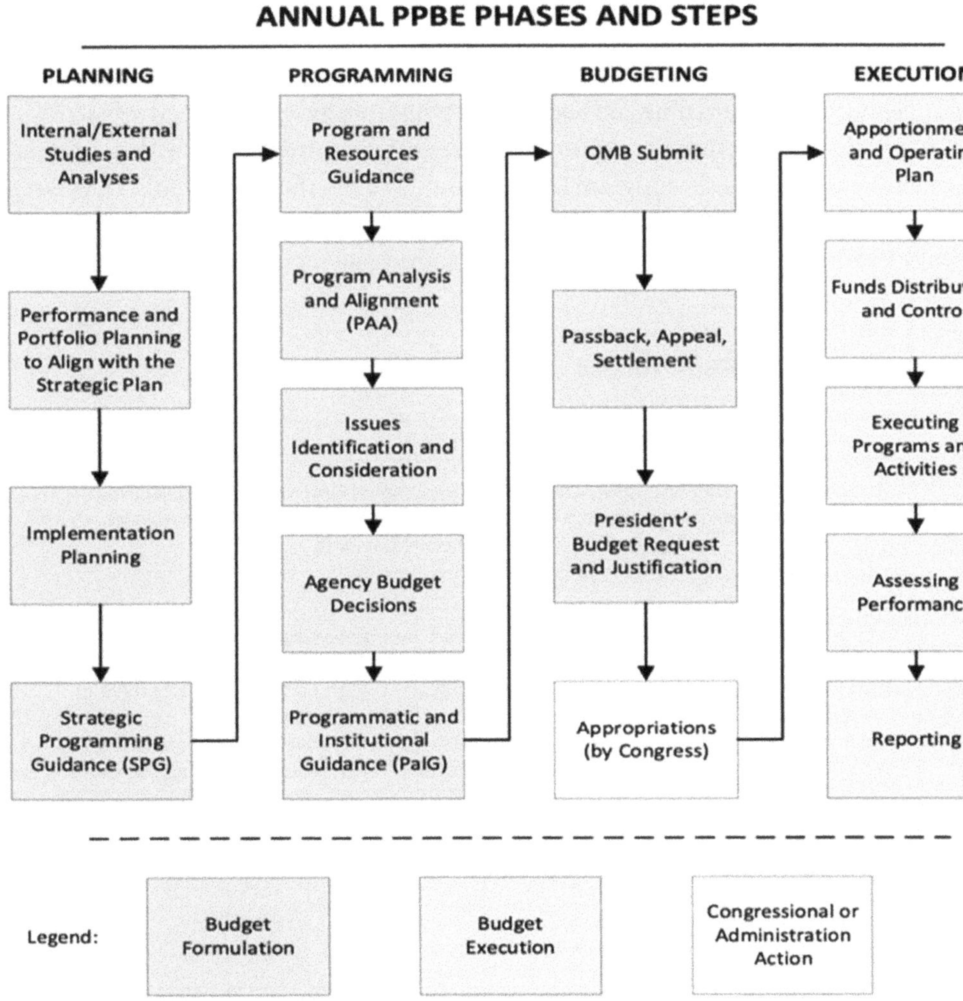

ANNUAL PPBE PHASES AND STEPS

PLANNING	PROGRAMMING	BUDGETING	EXECUTION
Internal/External Studies and Analyses	Program and Resources Guidance	OMB Submit	Apportionment and Operating Plan
Performance and Portfolio Planning to Align with the Strategic Plan	Program Analysis and Alignment (PAA)	Passback, Appeal, Settlement	Funds Distribution and Control
Implementation Planning	Issues Identification and Consideration	President's Budget Request and Justification	Executing Programs and Activities
	Agency Budget Decisions		Assessing Performance
Strategic Programming Guidance (SPG)	Programmatic and Institutional Guidance (PaIG)	Appropriations (by Congress)	Reporting

Legend:

Budget Formulation	Budget Execution	Congressional or Administration Action

SOURCE: Reproduced from NPR 9420.1A, 2021, p. 7.

a final request to Congress.[13] During the execution phase, NASA develops and maintains its congressionally approved budget operating plans and executes its appropriated programs.[14]

NASA's PA&E, established during the 2001–2005 tenure of NASA Administrator Sean O'Keefe, was the agency's first step toward a repeatable budget formulation process. This

---

[13] NPR 9420.1A, 2021.

[14] NPR 9420.1A, 2021.

office was the predecessor of SID and the originator of NASA's PPBE approach.[15] The resulting process, outlined in Figure 4.6, is carried out annually and over a series of years.

Although the phases of NASA's PPBE process are sequential and part of a cyclical system, planning and execution activities are continual. In other words, the process has concurrent phases focused on different time periods. For example, the strategic plan produced by SID covers a five-year period, but programming and budgeting activities are focused on single years.

NASA's budget is structured by mission, theme, program, and project, with appropriations generally allocated at the program level. Figure 4.7 shows the budget for NASA's Deep Space Exploration Systems initiative, which spans four themes, under which are programs and projects. Although it is rare for Congress to appropriate money at the project level, it

FIGURE 4.6
Flowchart of NASA's PPBE Process

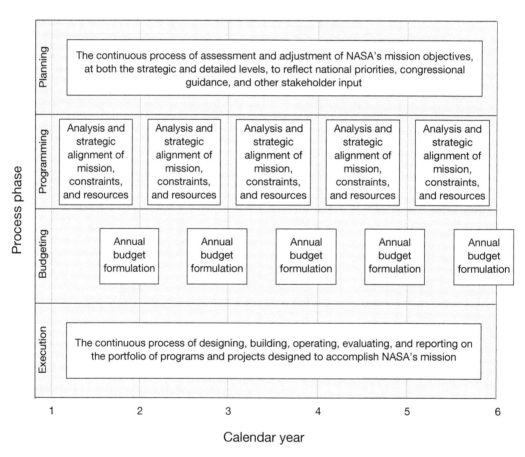

SOURCE: Adapted from NPD 1000.0C, 2020, pp. 30–31, Figures 4.0-2a and 4.0-2b.

---

[15] For more, see NASA, "NASA Establishes Office of Program Analysis and Evaluation," press release, June 20, 2005.

FIGURE 4.7

NASA's Budget Structure: Deep Space Exploration Systems Example

				Fiscal Year			
Budget Authority ($ in millions)	Op Plan 2021	Request 2022	Request 2023	2024	2025	2026	2027
NASA Total	**23,271.3**	**24,801.5**	**25,973.8**	**26,493.4**	**27,023.3**	**27,563.7**	**28,114.8**
Deep Space Exploration Systems	**6,396.5**	**6,750.2**	**7,478.3**	**6,798.1**	**6,813.4**	**6,912.3**	**7,040.8**
Common Exploration Systems Development	**4,538.7**	**4,483.7**	**4,668.4**	**3,613.8**	**3,111.9**	**2,845.6**	**2,376.8**
Orion Program	1,403.7	1,406.7	1,338.7	415.0	116.5	52.0	19.0
Crew Vehicle Development	1,387.8	1,388.8	1,325.3	415.0	116.5	52.0	19.0
Orion Program Integration and Support	15.9	17.9	13.4	--	--	--	--
Space Launch System	2,555.0	2,487.0	2,579.8	2,534.6	2,484.6	2,326.5	1,925.5
Launch Vehicle Development	2,488.4	2,414.0	2,505.7	2,455.8	2,393.2	2,234.4	1,832.7
SLS Program Integration and Support	66.6	73.0	74.0	78.8	91.3	92.0	92.7
Exploration Ground Systems	580.0	590.0	749.9	664.2	510.8	467.2	432.3
Exploration Ground Systems Development	569.2	585.3	747.3	660.0	503.8	460.0	425.0
EGS Program Integration and Support	10.8	4.7	2.6	4.2	7.0	7.2	7.3
Artemis Campaign Development	**1,672.1**	**2,062.0**	**2,600.3**	**2,973.8**	**3,489.3**	**3,853.9**	**4,450.7**
Gateway	501.5	685.0	779.2	754.5	685.4	661.7	758.2
Adv Cislunar and Surface Capabilities	45.0	82.0	59.6	57.8	62.9	425.2	555.6
Human Landing System	928.3	1,195.0	1,485.6	1,863.8	2,246.1	2,168.2	2,537.9
xEVA and Human Surface Mobility Program	197.3	100.0	275.9	297.7	494.9	598.8	599.1
Human Exp Requirements & Architecture	**9.5**	**9.5**	**48.3**	**48.9**	**49.5**	**50.0**	**50.5**
Moon & Mars Architecture	9.5	9.5	48.3	48.9	49.5	50.0	50.5
Mars Campaign Development	**176.2**	**195.0**	**161.3**	**161.6**	**162.7**	**162.7**	**162.8**
Exploration Capabilities	176.2	195.0	161.3	161.6	162.7	162.7	162.8

SOURCE: Reproduced from NASA, 2022c, p. BUD-1.
NOTE: Adv = Advanced; EGS = Exploration Ground Systems; Exp = exploration; Op = operations; SLS = Space Launch System; xEVA = exploration extravehicular activity.

does happen. For example, the two projects under the Orion Program in Figure 4.7 (Crew Vehicle Development and Orion Program Integration and Support) were funded directly by Congress.

Once NASA receives its appropriations (generally for programs), OCFO provides those resources to the mission directorates and to the Mission Support Directorate, which allocates funds to programs and projects at the geographically dispersed centers. There are two exceptions to this rule. The mission directorates transfer some funds into two accounts controlled by the Mission Support Directorate: SSMS and CECR. The SSMS account covers work that is not specific to any given center's programs or projects (e.g., security officers). The CECR account covers all construction work related to center programs and projects. NASA implemented this budget formulation process during the 2001–2005 tenure of former NASA Administrator Sean O'Keefe, who had adapted DoD's budget structure to suit NASA's needs.[16]

---

[16] NASA subject-matter experts, interviews with the authors, October 2022.

Decisionmakers and Stakeholders

Four senior leadership councils govern NASA by providing high-level oversight, setting requirements and strategic priorities, and guiding key agency assessments. Each council has a distinct charter and responsibilities.[17] Table 4.1 describes each council's role in the agency's decisionmaking.

NASA's PPBE process is marked by decentralized budget formulation in which its mission directorates consolidate budget inputs from the centers and then push the consolidated requests upward to the OCFO and Executive Council for final adjudication and approval. (All four of the management councils play a role in NASA'S PPBE process, but the Executive Council plays the most prominent role.) NASA also has a collaborative relationship with Congress, and congressional inputs are incorporated throughout NASA's PPBE process.[18] For example, NASA considers congressional guidance when defining strategic objectives in the planning phase, and subsequent requests for program resources are determined, in part, through engagements with both Congress and OMB in the budgeting phase. In the execution phase, NASA continuously monitors planning, programming, and budgeting data, and it

TABLE 4.1

NASA Management Councils

Name and Chair	Responsibilities
Executive Council *Chair: NASA Administrator*	• Determines NASA's strategic direction • Assesses NASA's progress toward achieving its vision • Serves as NASA's senior decisionmaking body for agency-wide decisions
Acquisition Strategy Council *Chair: NASA Associate Administrator*	• Provides high-level guidance to the Executive Council to inform the formulation of strategic programming guidance (SPG) • Approves acquisition approaches for large, high-profile programs as recommended by sponsoring mission directorates • Determines centers' work assignments and updates center roles • Evaluates mission needs and workforce capacity
Agency Program Management Council *Chair: NASA Associate Administrator*	• Serves as senior decisionmaking body for NASA's integrated mission portfolio • Assesses performance of NASA projects, programs, mission directorate portfolios, and integrated agency portfolio to ensure execution of strategic goals
Mission Support Council *Chair: NASA Deputy Associate Administrator for Business Operations*	• Serves as senior decisionmaking body regarding NASA's integrated mission support portfolio, mission support plans, and implementation strategies (including facility, infrastructure, workforce, and associated investments) • Determines and assesses mission support requirements

SOURCE: Adapted from NPD 1000.0C, 2020, p. 11, Table A.

---

[17] NPD 1000.0C, 2020.

[18] NASA subject-matter experts, interviews with the authors, August–November 2022.

issues reports to Congress on anticipated results and any planned corrective actions at various points in the fiscal year.[19]

Planning and Programming

As mentioned, OCFO manages NASA's budget formulation process, which NASA views as spanning the first three phases of PPBE (the blue boxes in Figure 4.5). Within OCFO, SID leads the planning and programming phases. In the planning phase, the SPG is developed and released. That guidance reflects the NASA Administrator's priorities and contains the top-line directorate budgets. The SPG is the first key budget guidance document produced, and it is usually published in February.[20]

NASA does not appear to have a process akin to DoD's JCIDS to generate requirements. Instead, it relies on a variety of sources. Requirements can be derived from presidential direction, the NASA Administrator's vision, congressional input, strategic plans, or decadal studies. Each decade, NASA and the National Research Council—the operating arm of the National Academies of Sciences, Engineering, and Medicine—collaborate to prioritize future research areas, observations, and notional missions.[21]

The second key document produced during this phase is the program and resources guidance (PRG), which the mission directorates provide to the centers. The PRG is usually published in April by mission directorate control account managers (CAMs). In essence, the SPG and PRG are collectively comparable to DoD's defense planning guidance. The mission directorate CAMs formulate and manage budget accounts across NASA's PPBE process using the SPG and then develop the PRG for center-level implementation.[22] In most cases, mission directorate–level resource management officers lead the PPBE process for top-line budget accounts identified in the SPG, while CAMs develop center-level guidance and brief NASA leadership on the center inputs.[23]

CAMs are also responsible for consolidating center-level requirements and conducting the program analysis and alignment to resolve conflicts within budget accounts. These consolidations and resolutions are briefed to the mission directorates in June or July. After that, the Executive Council makes final budgeting decisions, which are codified in programmatic and institutional guidance.

---

[19] NPD 1000.0C, 2020.

[20] NASA subject-matter experts, interviews with the authors, September 2022.

[21] Decadal studies happen more frequently than once every ten years because they are specific to mission directorates (e.g., Science, Space). Decadal documents were published in 2010, 2013, 2018, and 2022. For more on recent decadal studies in NASA's Science Mission Directorate, see NASA Science, "Most Recent Decadal Studies," webpage, March 9, 2023.

[22] NPR 9420.1A, 2021.

[23] NPR 9420.1A, 2021.

During the planning phase, centers develop required documents and internal guidance, such as strategic workforce plans, that are pushed up to the mission directorates for consolidation. Because centers do not exercise programmatic authority over projects, they communicate and collaborate with mission directorates on budget matters.[24] Budget requests are typically reconciled at the mission directorate level before they are submitted to the Executive Council, which is chaired by the NASA Administrator. After Executive Council approval, NASA headquarters submits the budget to OMB and Congress.

SID establishes and maintains cost and schedule estimates. It documents these estimates in program or project plans and validates the estimates at both the program and mission directorate levels.[25] NASA uses several information technology tools to support this portion of the PPBE process. It uses the Strategic Management System to assess resource allocation and planned performance goals.[26] N2, NASA's budget formulation suite, provides comprehensive data on budgets and workforces across the enterprise. This database allows users with specified roles to view and revise budget inputs as needed and to create an audit report that tracks every entry into the database by project, username, date, and budget amount changed. Another tool, eBudget, also supports budget formulation and the PPBE process more broadly, including the development and submission of a congressional budget justification document. eBudget allows NASA to integrate financial data across the entire agency.[27]

Budgeting and Execution

OCFO's Budget Division leads the budgeting phase and the development of the congressional justification, including NASA's input for the President's Budget request. Congress appropriates NASA funding across only eight budget accounts: six mission directorates (including the Mission Support Directorate, which manages the SSMS and CECR accounts); the Science, Technology, Engineering, and Mathematics (STEM) Engagement office; and NASA's OIG. Figure 4.8 shows NASA's budget appropriations, from FY 2004 (the year before the agency implemented its PPBE process) to its FY 2023 request.

In the congressional justification that OCFO submits to OMB and Congress, NASA breaks its budget down by mission, theme, program, and, occasionally, project (as shown in Figure 4.4). Each congressional justification includes approximately 155 budget line items spread across the nine budget accounts. Figure 4.9 shows the nine mission directorate budget lines from FY 2015 to FY 2023. In FY 2022, the median and mean of these budget lines were $1.10 billion and $2.67 billion, respectively. It is notable that NASA's funds are not broken out

---

[24] NPD 1000.0C, 2020, p. 21.

[25] This insight was primarily gleaned through semistructured interviews with NASA subject-matter experts, as well as our review of NPD 1000.0C, 2020, p. 23.

[26] NPD 1000.0C, 2020, p. 36.

[27] NASA OIG, *Integrated Financial Management Program Budget Formulation Module*, Audit Report IG-04-017, March 30, 2004.

FIGURE 4.8

NASA's Budget, FYs 2004–2023

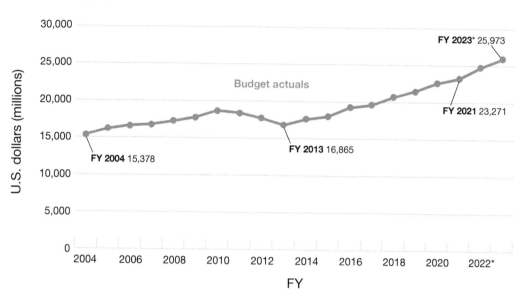

SOURCES: Authors' analysis of fiscal year budget request summary data from NASA, "Previous Years' Budget Requests," webpage, last updated February 27, 2023b. Also see NASA, 2022c.
NOTE: Actuals were derived from subsequent President's Budget request documents. Values for FYs 2022–2023 are reported as requests and denoted with an asterisk (*).

by the equivalent of DoD appropriation categories (e.g., RDT&E; procurement; O&M; military personnel; and military construction).[28]

After NASA receives its annual appropriation, it uses its congressional operating plan (COP), agency operating plan (AOP), and agency execution plan (AEP) to ensure that appropriated funds are used in compliance with their intent and congressional mandates.[29] The COP is submitted to Congress and sets forth a high-level plan for how NASA intends to spend its budget during the fiscal year. The AOP is an internal plan, based on the COP, that provides greater detail and includes all programs and projects. The AEP is a detailed financial plan, based on the AOP, that represents the agency's total expected budget authority in the execution year, establishes planning controls for programs and projects, and sets forth how funds will be distributed.

The AOP is updated two or three times annually when new requirements become known. If the changes exceed limitations established in the COP, NASA must submit a new plan to Congress. Whenever the COP or AOP is updated, NASA revises the AEP to ensure that funds are used appropriately. Mission directorates provide budgetary inputs for the COP and AOP,

---

[28] This insight was derived from multiple semistructured interviews with NASA subject-matter experts and our analysis of budget request documents (see NASA, 2023b).

[29] NPR 9470.1, 2008, chapter 2.

FIGURE 4.9
NASA's Budget Lines, FYs 2015–2023

Budget Line	FY 2015 Actual	FY 2016 Actual	FY 2017 Actual	FY 2018 Actual	FY 2019 Actual	FY 2020 Actual	FY 2021 Enacted	FY 2022 Enacted (3/15/22)	FY 2023 President's Budget Request
Deep Space Exploration Systems	$3,542.7	$3,996.2	$4,184.0	$4,790.0	$5,044.8	$5,959.8	$6,555.4	$6,791.7	$7,123.2
Space Operations	$4,625.5	$5,032.3	$4,942.5	$4,749.2	$4,640.4	$1,100.0	$3,988.2	$4,041.3	$4,621.4
Space Technology	$600.3	$686.4	$826.5	$760.0	$926.9	$4,134.7	$1,100.0	$1,100.0	$1,437.9
Science	$5,243.0	$5,584.1	$5,762.2	$6,211.5	$6,886.6	$7,143.1	$7,301.0	$7,614.4	$7,988.3
Aeronautics	$642.0	$633.8	$656.0	$690.0	$724.8	$783.9	$828.7	$880.7	$971.5
STEM Engagement	$119.0	$115.0	$100.0	$100.0	$110.0	$120.0	$127.0	$137.0	$150.1
SMSS	$2,754.6	$2,772.4	$2,768.6	$2,826.9	$2,755.0	$2,913.3	$2,936.5	$3,020.6	$3,208.7
CECR	$446.1	$427.4	$375.6	$569.5	$372.2	$432.5	$390.3	$410.3	$424.3
Inspector General	$37.0	$37.4	$37.9	$39.0	$39.3	$41.7	$44.2	$45.3	$48.4
Total NASA	**$18,010.2**	**$19,285.0**	**$19,653.3**	**$20,736.1**	**$21,500.0**	**$22,629.0**	**$23,271.3**	**$24,041.3**	**$25,973.8**

$ millions	FY 2022 Enacted (3/15/22)
Median	$1,100.0
Mean	$2,671.3

SOURCE: Reproduced from NASA subject-matter expert, response to follow-up questions, Microsoft Excel spreadsheet provided to the authors, December 2022.

while the CAMs use the AEP to distribute funding across programs and projects at the center level.[30]

With NASA's budget typically allocated at the program level or above, its reprogramming requirements differ significantly from DoD's.[31] NASA is required to submit reprogramming documents (i.e., an updated COP) when a budget account changes by $500,000.[32] Thus, although NASA's $500,000 reprogramming threshold is significantly lower than DoD's $10 million threshold, having the funds appropriated at a higher level than would be normal for DoD still appears to provide NASA with significant flexibility.[33]

As mentioned, the Strategic Management System gives NASA integrated, near-real-time access to planning, budgeting, and programmatic data to enable timely decisionmaking. However, it does not appear that NASA has the capability to assess the effectiveness of spending after the year of execution. NASA's PA&E reportedly had studies, analysis, and independent assessment functions prior to SID's creation. However, these functions do not appear to have been transferred to SID, and, as of this writing, there does not appear to be an office in the agency with the requisite manpower and skill sets to execute large-scale studies of the effectiveness of NASA's prior spending.[34]

While SID may not evaluate the effectiveness of NASA's past spending, it is responsible for developing NASA's strategic plan, which looks forward five years. These strategic plans consider all NASA budget accounts that—with the exception of those for the OIG and CECR—have two-year durations.[35] This approach is different from that used in DoD. These two-year funds allow NASA to better plan for continuing resolutions. According to our interviewees, NASA has a 90- to 95-percent obligation goal in the first year, affording the agency the flexibility to use the remainder in the next fiscal year to cover obligations until a budget is passed. In other words, this approach allows NASA to forward-fund contracts. Mission directorate resource management officers continually monitor funding execution. Coupled with priori-

---

[30] NPR 9470.1, 2008.

[31] For more on DoD's reprogramming requirements, see Office of the Under Secretary of Defense (Comptroller), "Summary of Reprogramming Requirements, Effective for FY2021 Appropriation," January 6, 2021.

[32] This insight was gleaned from multiple interviews with NASA subject-matter experts. NASA's reprogramming threshold is approximately 0.05 percent of the median budget line item.

[33] The highest level of appropriation within NASA is at the mission directorate level. From there, funds are disbursed among themes, programs, and projects that may cut across multiple centers. The equivalent DoD appropriation level would be the service level. Additionally, NASA's reprogramming authority is proportionately about twice as high as DoD's, assuming that their overall budgets are approximately $26 billion and $840 billion, respectively.

[34] NASA subject-matter experts, interviews with the authors, September–October 2022.

[35] NASA subject-matter experts, interviews with the authors, August–October 2022.

tized, essential activities in OCFO's contingency plan,[36] this allows leadership to assess risk, develop mitigation plans, and continue to execute activities during a continuing resolution.[37]

Oversight

Congressional oversight of NASA and its budget is exercised by several committees (House and Senate Appropriations and Budget Committees) and subcommittees (House and Senate Subcommittees on Commerce, Justice, Science, and Related Agencies).[38] NASA's budget process typically begins in February and follows a timeline similar to DoD's, with three years in play at any given time: execution (current year), budget request (next year), and planning (two years in advance). NASA's process has an underlying assumption, however: The timing of the process assumes that Congress will pass the budget on time.

NPR 9470.1 outlines the financial management requirements for budget execution. These requirements apply to all NASA employees and facilities, including NASA headquarters and centers; its federally funded research and development center, the Jet Propulsion Laboratory; and its contractors, grant recipients, and other parties to agreements.[39]

NASA uses its COP, AOP, and AEP to ensure that its appropriated funds are used in compliance with their intent and congressional mandates.[40] As mentioned, the COP is submitted to Congress and sets forth a high-level plan for how NASA intends to spend its budget during the fiscal year. The AOP is an internal plan, is updated two or three times per year if new requirements become known, and includes all programs and projects. These documents are created by the OCFO with the approval of the Administrator. The AEP is a detailed financial plan that represents the total budget authority that NASA expects during the execution year, establishes planning controls for programs and projects, and sets forth how funds will be distributed.

NASA publishes a report as required by the Good Accounting Obligation in Government Act of 2019.[41] The most recent version of this report that was available at the time of this research was dated January 31, 2022. Additionally, NASA produces an annual financial report that contains a "Summary of Financial Statement Audit and Management Assurances."[42] This summary identifies material challenges (if any) in the prior year and their status. NASA

---

[36] NPR 9470.1, 2008, chapter 3.

[37] NASA subject-matter experts, interviews with the authors, September and November 2022.

[38] NASA OCFO, "Congressional Committees," webpage, last updated March 25, 2019.

[39] NPR 9470.1, 2008. This policy is based on laws, as outlined in the document. At the time of this research, NPR 9470.1 was set to expire on December 24, 2023, and it was unclear whether or when the document would be renewed or succeeded by an updated NPR.

[40] NPR 9470.1, 2008, chapter 2.

[41] Public Law 115-414, Good Accounting Obligation in Government Act of 2019, January 3, 2019.

[42] NASA, 2022b.

also complies with OMB's requirement for an annual data call, mandated by the Payment Integrity Information Act of 2019.[43]

As of FY 2022, NASA had received 12 consecutive clean opinions from an external auditor. The unmodified opinion was that NASA's financial statements presented its financial position and operations results fairly and conformed to accepted accounting principles for federal agencies. NASA produces a financial report as part of its annual *Volume of Integrated Performance*, which summarizes past-year performance and provides updates on the current year's performance plan. The report is published in conjunction with the President's Budget request.[44]

Analysis of NASA's Budgeting Process

Drawing from our analysis of NASA's budgeting and resource allocation processes, we identified the strengths and challenges presented in Table 4.2. In the following sections, we discuss these strengths and challenges, along with potential lessons for DoD.

Strengths

NASA's Culture of Collegiality Is Embedded in Its Resource Management Process

One of NASA's defined attributes (see Figure 4.1) is its team orientation: "Come together as one to solve complex issues. Innovation is a staple, teamwork is a must and everyone's opinion counts."[45] In our interviews of personnel from NASA headquarters, directorates, and centers,

TABLE 4.2

Strengths and Challenges of NASA's PPBE Process

Strengths	Challenges
• NASA's culture of collegiality is embedded in its resource management process. • Appropriations by mission, theme, and program provide flexibility. • Two-year funding helps NASA function in times of uncertainty (such as when there are continuing resolutions). • NASA's funds are appropriated differently than DoD's, providing execution flexibility. • NASA's requirement process is less formal than DoD's JCIDS.	• OCFO oversees the planning, programming, and budgeting process. • SID is not analogous to CAPE in the OSD. • SSMS and CECR funds are not directly allocated to the Mission Support Directorate.

---

[43] Public Law 116-117, Payment Integrity Information Act of 2019, March 2, 2020.

[44] NASA, "NASA Receives 12th Successive 'Clean' Financial Audit Rating," press release, November 15, 2022g.

[45] NASA, 2022e.

it appeared that colleagues at all echelons worked as a highly functioning team to support the PPBE process.[46] Roles and responsibilities were well understood, information flows were documented, and there were multiple opportunities for discussion (e.g., about the development and implementation of the SPG, program analysis and alignment, and programmatic and institutional guidance). Disconnects were resolved primarily at the mission directorate level, and we were told that it was rare for issues to be elevated to the Executive Council for the most senior leaders to resolve.

Appropriations by Mission, Theme, and Program Provide Flexibility

NASA's budget is developed at the highest level of mission, then broken down into successively lower levels of theme, program, and project. According to conversations with OCFO members and as confirmed by NASA's FY 2023 congressional budget justification, the majority of NASA's funding is requested above the project level.[47] These higher-level requests allow the mission directorates to move funds to selected projects within themes and programs without further congressional approval. Additionally, NASA updates its COP two or three times per year to reflect real-world changes; these updates to Congress are essentially NASA's version of DoD's omnibus reprogramming request.

Two-Year Funding Helps NASA Function in Times of Uncertainty

According to our interviewees, all of NASA's appropriations, except for OIG and CECR, have two-year durations.[48] Additionally, NASA's goals for obligations and expenditures allow funds to carry over into the second year to act as a cushion for continuing resolutions.[49] With obligation goals in the 90- to 95-percent range for the first year, NASA can have funds available at the start of the second year to keep programs on schedule.

NASA's Funds Are Appropriated Differently Than DoD's

Our review of NASA's FY 2023 congressional justification documents indicated that NASA does not request, nor is it funded with, appropriations split into such categories as RDT&E, procurement, and O&M in the same manner as DoD, and this finding was confirmed during our interviews.[50] Therefore, NASA does not appear to have the same types of restrictions as DoD with respect to using specific funding for specific activities (e.g., using RDT&E only during the design and development stages of a program).

---

[46] NASA subject-matter experts, interviews with the authors, August–November 2022.

[47] NASA subject-matter experts, interviews with the authors, September 2022; NASA, 2022c.

[48] NASA subject-matter experts, interviews with the authors, August–November 2022.

[49] Although carryover funding may mitigate some aspects of continuing resolutions, this is not its primary purpose. Carryover funding is designed to counter the use-it-or-lose-it mentality and behavior associated with one-year funding.

[50] NASA, 2022c; NASA subject-matter experts, interviews with the authors, August–November 2022.

NASA's Requirement Process Is Less Formal Than DoD's JCIDS

NASA's budgeting requirements come primarily from decadal studies, presidential and administrator priorities, and emerging advancements in technology instead of from the threat-driven requirements or capability gaps that DoD addresses through JCIDS. According to our interviewees, inputs from NASA's decadal studies provide a deeper understanding of technological advances so that the agency can account for them in its budget process.[51]

Challenges

OCFO Oversees the Planning, Programming, and Budgeting Process

Because the planning, programming, and budgeting are handled by one NASA organization, there is a possibility that conflicts of interest might arise. This structure makes it difficult to conduct an independent examination of the budget. For example, in contrast with DoD's CAPE, NASA's SID, which develops strategic planning guidance that informs programming and budgeting phases, is situated within OCFO.

SID Is Not Analogous to CAPE

SID is housed within OCFO and, therefore, may not be considered an independent organization when it scrutinizes budget submissions from the centers and the directorates en route to NASA headquarters. Interviewees noted that independent agencies may be able to identify overruns faster than NASA's current evaluation-oriented offices could, primarily because of workforce capacity challenges.[52]

In contrast, DoD's CAPE is separate from the services and defense agencies. It was created by the Weapon Systems Acquisition Reform Act of 2009, with its implementation formalized in Directive-Type Memorandum 09-027, Implementation of the Weapons Systems Acquisition Reform Act of 2009.[53] CAPE's goals and principles include bringing discipline to DoD's PPBE process, providing deep insight into the costs of major acquisition programs, and conducting "fact-based independent and objective analyses even when it is uncomfortable."[54]

---

[51] NASA subject-matter experts, interviews with the authors, August–September 2022; NASA Science, 2023. Decadal studies occur more frequently than every ten years. NASA asks the National Research Council to look out ten or more years and prioritize areas that might merit observation or notional missions to make those observations.

[52] NASA subject-matter experts, interviews with the authors, September 2022.

[53] Public Law 111-23, Weapon Systems Acquisition Reform Act of 2009, May 22, 2009; Directive-Type Memorandum 09-027, Implementation of the Weapon Systems Acquisition Reform Act of 2009, Office of the Under Secretary of Defense for Acquisition, Technology and Logistics, U.S. Department of Defense, December 4, 2009, incorporating change 1, October 21, 2010. CAPE originated as the Office of Systems Analysis in 1961, which transitioned to the Office of Program Evaluation and Analysis in the 1970s (OSD, Cost Assessment and Program Evaluation, homepage, undated).

[54] OSD, Cost Assessment and Program Evaluation, undated.

Although overruns still occur in DoD, it is possible that they would be worse without the independent oversight of an analytically driven organization.

SSMS and CECR Funds Are Not Directly Allocated to the Mission Support Directorate

SSMS and CECR funds are provided to the mission directorates, which then reallocate them to the Mission Support Directorate for non–program-specific (SSMS) or construction (CECR) work at the centers. This reallocation appears to be an additional step, which increases bureaucracy for an indeterminate benefit.

Applicability

Because of the size difference between NASA and DoD—in terms of the number of people and organizations, as well as levels of funding—it is difficult to be sure which of NASA's strengths could be scaled for adoption by DoD. However, if such scaling were feasible, DoD might take a more collegial approach to the PPBE process by encouraging budgetary planning *across* its organizations. If the various DoD organizations did not view the budget process as a zero-sum game, with each organization competing for funding against others, greater cooperation and reduced duplication of capability might be possible.

Given DoD's complexity and the significant portion of the federal government's budget it consumes, it warrants a greater degree of legislative oversight. This oversight is sometimes exercised through appropriations bills (through which funds are allocated). However, a difference between DoD and NASA is reflected in how funds are appropriated to NASA and for what purposes those appropriated funds can be used. For DoD, its current level of control can have unintended consequences (e.g., frenzied end-of-fiscal-year spending before funds expire, limited ability to direct funds to priorities). Although NASA and DoD are both overseen by subcommittees of the House and Senate appropriations committees, the funds for NASA appear to be allocated at a higher level, providing NASA with more flexibility to adjust in response to changing priorities or real-world events.

Lessons from NASA's Budgeting Process

Lesson 1: A Culture of Cooperation Enhances Overall Execution

NASA's more collegial approach to PPBE could reduce interdepartmental competition for funding in DoD and improve cooperation across the department's various components. A greater collective focus on common goals could improve how DoD "provides the military forces needed to deter war and ensure [the] nation's security."[55]

---

[55] DoD, "About," webpage, undated.

Lesson 2: A Less Restrictive Appropriation Structure Improves Flexibility

NASA requests and is allocated funding differently than DoD. Because NASA's funds are primarily appropriated at the mission, theme, and project levels to mission directorates, NASA has some flexibility to align project funding to meet changing priorities or real-world circumstances.

Lesson 3: Independent Assessment Could Inform NASA Leadership

Having an independent organization assess the PPBE and report to senior decisionmakers could be a positive development for NASA. In DoD, this is CAPE's mission, and it has the support of senior leadership and the statutory authority to execute this mission. DoD organizations understand the need for such an entity and already provide information that CAPE needs for its assessments, which inform senior decisionmakers about resource allocation and cost estimation.

Lesson 4: Variable Obligation Goals Can Help Mitigate the Impact of Continuing Resolutions

NASA has obligation goals of 90 to 95 percent in the first year of two-year funds. These goals allow for some funding to be obligated in the second year, typically at the start of the fiscal year. Because continuing resolutions are a real possibility, this carryover funding can mitigate many funding shortfalls that might result at the start of a fiscal year.

Lesson 5: Clean Audits Imbue Oversight Organizations with Trust

For the past 12 years, NASA has had 12 consecutive clean opinions from an external auditor. These opinions can reassure Congress that NASA spends its funding in accordance with congressional direction and conforms to accepted accounting principles for federal agencies. With this level of independent assurance, NASA continues to foster a positive relationship with relevant congressional committees and subcommittees, which could account for some of the increased flexibility in its PPBE process. If DoD were able to consistently achieve a clean audit, that might translate to increased flexibility for it as well.

Table 4.3 summarizes these five lessons.

TABLE 4.3

Summary of Lessons from NASA's Budgeting Process

Theme	Lesson Learned	Description
Planning and programming	Lesson 1: A culture of cooperation enhances overall execution.	A collegial approach appears to reduce internal conflict and improve cooperation among otherwise competing internal organizations.
Planning and programming	Lesson 2: A less restrictive appropriation structure improves flexibility.	NASA's funding is appropriated at a higher level and with apparently fewer categories than DoD's.
Budgeting and execution	Lesson 3: Independent assessment could inform NASA leadership.	An assessment by a structurally independent organization helps ensure that NASA's senior leadership is well informed about resource allocation and cost estimation.
Budgeting and execution	Lesson 4: Variable obligation goals can mitigate the impact of continuing resolutions.	NASA has 90- to 95-percent obligation goals for the first year of two-year funds, allowing some funds to carry over into the second year.
Oversight	Lesson 5: Clean audits imbue over-sight organizations with trust.	NASA's 12 consecutive years of clean audits can reassure Congress that it is properly handling the taxpayers' money.

Office of the Director of National Intelligence

Anthony Vassalo and Sarah W. Denton

After the terrorist attacks of September 11, 2001, intelligence integration became a principal concern for the federal government. IRTPA was designed to promote the operational integration of the IC through the creation of ODNI as IC coordinator and to improve information-sharing across the IC.[1] Prior to IRTPA, the director of the CIA had been dual-hatted, also acting as the nominal head of the IC, with the assistance of a small community of management staff within CIA headquarters.[2]

ODNI began operations in 2005. Its early years were hampered by instability in staffing the DNI and PDDNI roles. From 2005 to 2010, the DNI position was filled by John Negroponte, Mike McConnell, Dennis Blair, and James Clapper. Clapper's longer tenure than his predecessors, from 2010 to 2017, provided ODNI with an opportunity to establish itself within the IC. Moreover, Clapper provided an overarching vision, rationale, and value proposition for the office's relationship with the IC and the concept of vertical and horizontal integration of the IC and its activities.[3]

Additionally, Clapper and his PDDNI, Stephanie O'Sullivan, built on the work that ODNI staff had begun under Dennis Blair to piece together the IC budget process, which would be codified in 2011 as ICD 116.[4] Clapper and O'Sullivan then used ICD 116 as a key component for integrating the budget processes across the IC.

ODNI's mission is to lead intelligence integration across the intelligence enterprise and forge an IC that gives policymakers and warfighters a decisionmaking advantage by consis-

---

[1] Pub. L. 108-458, 2004; Rosenwasser, 2021. This view of ODNI's role was also shared by an ODNI subject-matter expert in an interview with the authors in February 2023.

[2] ODNI subject-matter experts, interviews with the authors, November–December 2022; Director of Central Intelligence Directive 3/3, 1995.

[3] ODNI subject-matter experts, interviews with the authors, November 2022–February 2023; Clapper and Brown, 2021.

[4] ICD 116, 2011.

tently and routinely delivering the most insightful intelligence possible.[5] The IC is a coalition of 18 agencies and organizations, including ODNI,[6] as follows:

- two independent organizations
  - ODNI
  - CIA
- nine DoD elements
  - Defense Intelligence Agency
  - National Security Agency
  - National Geospatial-Intelligence Agency
  - National Reconnaissance Office
  - U.S. Air Force
  - U.S. Army
  - U.S. Marine Corps
  - U.S. Navy
  - U.S. Space Force
- seven elements of other departments and agencies
  - U.S. Coast Guard Intelligence
  - U.S. Department of Energy, Office of Intelligence and Counterintelligence
  - DHS, Intelligence and Analysis
  - U.S. Department of Justice, Drug Enforcement Agency, Office of National Security Intelligence
  - U.S. Department of Justice, FBI
  - U.S. Department of State, Bureau of Intelligence and Research
  - U.S. Department of Treasury, Office of Intelligence and Analysis.

Figure 5.1 depicts all departments and agencies that make up the IC, with ODNI in the center. IRTPA gives the DNI primary responsibility for the following activities:[7]

- serving as the head of the IC
- acting as principal adviser for intelligence matters related to national security
- directing the execution of the NIP, developing an annual consolidated budget for the NIP, and managing the NIP's appropriated funds
- participating in the development of the annual budget for the MIP.[8]

---

[5] ODNI, undated-d.

[6] ODNI, undated-g.

[7] Pub. L. 108-458, 2004.

[8] The MIP funds defense intelligence activities intended to support tactical military requirements and operations. Prior to 2005, military-specific intelligence activities were not covered in the National Foreign Intelligence Program (NFIP) and were known as Tactical Intelligence and Related Activities

FIGURE 5.1

U.S. Intelligence Community Elements

SOURCE: Reproduced from ODNI, undated-g.

(TIARA). In 2005, the Secretary of Defense signed a memorandum that combined TIARA and the Joint Military Intelligence Program (JMIP) to form the MIP. For more information, see Michael E. DeVine, *Defense Primer: Budgeting for National and Defense Intelligence*, IF10524, Congressional Research Service, November 17, 2022.

The DNI is authorized to perform the following budgetary functions:[9]

- provide guidance to IC components regarding the development of the NIP based on intelligence priorities established by the President
- develop and determine an annual consolidated NIP budget
- present a finalized budget for presidential approval
- transfer and reprogram funds within the NIP with the approval of the director of OMB and in consultation with the affected agencies.

Since 2007, the number of military and government civilian employees at ODNI has remained relatively consistent. Of the office's approximately 1,700 personnel, 40 percent are on rotation from the other 17 IC elements.[10] The National Counterterrorism Center, National Counterproliferation and Biosecurity Center, National Counterintelligence and Security Center, and Foreign Malign Influence Center are all part of ODNI.[11] Unlike other headquarters elements, such as those within DoD, ODNI employs fewer contractors than government civilian and military staff.[12]

The NIP and the MIP are the two major components of the U.S. intelligence budget. The NIP "includes all programs, projects, and activities of the intelligence community as well as any other intelligence community programs designated jointly by the DNI and the head of department or agency, or the DNI and the President."[13] In contrast, the MIP is "devoted to intelligence activity conducted by the military departments and agencies in the Department of Defense that support tactical U.S. military operations."[14] The MIP is managed through the standard DoD PPBE process by the Under Secretary of Defense for Intelligence and Security.[15]

Figure 5.2 shows NIP and MIP appropriations since FY 2006, along with total IC budget appropriations. ODNI's most recent appropriation was for $89.8 billion in FY 2022. In general, the NIP has accounted for roughly two-thirds to three-quarters of the IC budget over time, while the MIP has accounted for about one-quarter to one-third.

ODNI manages the NIP through the IPPBE System (also the title of ICD 116). The IPPBE process was specifically modeled on DoD's PPBE process, with some modifications to ensure that the IPPBE suited ODNI's mission. This PPBE adaptation was done both by necessity, to

---

[9] Pub. L. 108-458, 2004.

[10] ODNI subject-matter experts, interviews with the authors, August 2022–February 2023; ODNI, 2017.

[11] ODNI subject-matter experts, interviews with the authors, November 2022–February 2023; Lowenthal, 2019, p. 42; ODNI, 2022.

[12] ODNI, 2017.

[13] ODNI, undated-f.

[14] ODNI, undated-f.

[15] DoDD 5105.12, *Director of Cost Assessment and Program Evaluation*, Office of the Chief Management Officer of the Department of Defense, August 14, 2020.

account for the presence of DoD elements in the IC, and by design, because the architects of the IPPBE process were well versed in PPBE and viewed it as the best model for ODNI's complex organizational structure.[16] One notable and deliberate difference between the IPPBE and PPBE processes is ODNI's substitution of *evaluation* for DoD's *execution*.[17] The authors of ICD 116 put in place a comprehensive evaluation system to help shape the future allocation of resources within the IPPBE process.[18]

The mechanisms of the ODNI budgeting process have evolved with changing leadership priorities, resourcing constraints, and staff reorganizations. However, the focus on *evaluation*, several other core themes, and the overarching process have remained constant since ICD 116 was published in 2011. This chapter covers several other core themes:

- The IPPBE process is perceived as a key *integrating* tool for the IC, providing an opportunity for horizontal integration, federated and shared investments, and discussions about risk mitigation and prioritization.

FIGURE 5.2

U.S. Intelligence Community Budget, FYs 2006–2022

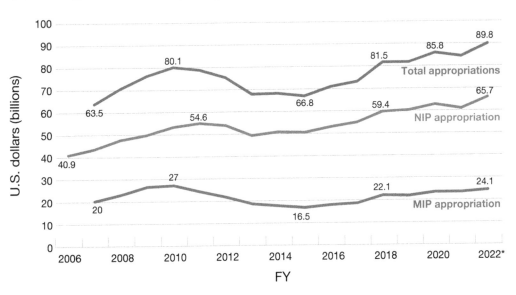

SOURCE: Features information from ODNI, undated-f.
NOTE: Actuals were derived from subsequent President's Budget request documents. Values for FY 2022 are reported as requests and denoted with an asterisk (*).

---

[16] ODNI subject-matter experts, interviews with the authors, August–September 2022.

[17] Note that the execution phase is incorporated into the budgeting phase. ODNI subject-matter experts, interviews with the authors, August–September 2022.

[18] ODNI subject-matter experts, interviews with the authors, August–September 2022. Also see ICD 116, 2011. The accountability and evaluation tools in the IPPBE process have changed over time but have included SERs, MISs, CIG compliance reports, and strategic program briefings.

- The IPPBE process provides a way to *synchronize* broader IC activities and community-wide outreach by the various components of ODNI.
- The IPPBE process provides various forums for establishing *transparency.*
- The IPPBE process *affirms* the role and position of the DNI in relation to the IC component heads.[19]

Overview of ODNI's Budgeting Process

ODNI's authorities over the formulation and management of the NIP, as well as the role of the DNI in DoD's management of the MIP, are codified in U.S. Code, Title 50, Section 3024 and established by Executive Order 12333.[20] ODNI's related authorities include the following:[21]

- providing guidance to IC elements for developing the NIP budget
- developing the annual consolidated NIP budget and presenting it to the President for approval
- ensuring effective execution of the annual budget for intelligence activities
- managing NIP appropriations by allocating them through the heads of IC agencies
- monitoring the NIP's implementation and execution, which may include audits and evaluations
- reprogramming NIP funds with prior DNI approval and notifying appropriate congressional committees of the intent to obligate or expend funds for intelligence-related activities that differ from those specifically authorized by Congress
- providing consultation to the Secretary of Defense on all MIP reprogramming requests
- eliminating waste and unnecessary duplication within the IC
- collaborating with the Under Secretary of Defense for Intelligence and Security to facilitate coordination of NIP and MIP budgets to satisfy national intelligence needs.

The NIP and MIP budgets are managed through separate but coordinated processes: IPPBE for the NIP and PPBE for the MIP. Both processes begin more than two years before the beginning of the budgeted fiscal year, and both are used to plan, program, and budget

---

[19] ODNI subject-matter experts, interviews with the authors, August 2022–February 2023. Also see ICD 116, 2011.

[20] U.S. Code, Title 50, Section 3024(c), Responsibilities and Authorities of the Director of National Intelligence, Budget Authorities; Executive Order 12333, *United States Intelligence Activities*, Executive Office of the President, December 4, 1981. Also see ICD 104, *National Intelligence Program (NIP) Budget Formulation and Justification, Execution, and Performance Evaluation*, Office of the Director of National Intelligence, April 30, 2013.

[21] See U.S. Code, Title 50, Section 3024, paras. (c)(1)(A), (c)(1)(B), (c)(1)(C), (c)(4), (c)(5)(A), (c)(5)(B), (c)(5)(C), (d)(1)(A), (d)(1)(B), (d), (f)(5), (h)(2), (p), (r); and U.S. Code, Title 50, Section 3094, Funding of Intelligence Activities, para. (a)(3)(C). These authorities are also broken out by section of U.S. Code in an unclassified document provided by our contact at ODNI.

project costs, manpower needs, and required capabilities for five years into the future (i.e., the FYDP).[22] Figure 5.3 offers a simplified look at the IPPBE process.

The IPPBE process has evolved over the years, but its timeline has remained fairly constant. As shown in Figure 5.3, planning for the execution year ("Fiscal Year" in the figure) is underway by March three years earlier FY(+3) and ends around the end of December in FY(+2). Roughly bracketing that planning phase, ODNI staff generate the MIL and the DNI's strategic priorities in May and December, respectively, and both feed directly into the programming phase. The programming phase, which overlaps significantly with the planning phase, begins around April or May of FY(+3) and ends around May or June of FY(+2). During the programming phase, outputs include MISs, the draft CIG, the strategic program briefing, and the final CIG, which feeds into the budgeting phase.

The budgeting phase begins around January or February in FY(+2) and runs through February of FY(+1). During this phase, ODNI staff produce procedural guidance, which informs

FIGURE 5.3
Simplified Flowchart of ODNI's IPPBE Process

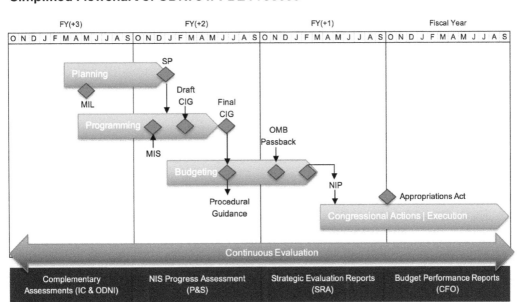

SOURCE: Adapted from an ODNI subject-matter expert, "IPPBE Process and Products," handout provided to the authors during interview, August 2022.

NOTES: For a more detailed view of the IPPBE process, see ODNI, "Intelligence Community Resource Process Guide," undated-c. Parenthetical FY designations (+3, +2, and +1) indicate the number of fiscal years of advance budget planning for the execution year ("Fiscal Year"). For example, FY(+3) indicates planning done three years prior to the execution year. MIL = major issues list; NIS = National Intelligence Strategy; P&S = Policy and Strategy Office; SP = strategic priorities; SRA = Systems and Resource Analysis Office.

---

[22] Robert A. Mirabello, "Guide to the Study of Intelligence: Budget and Resource Management," *The Intelligencer: Journal of U.S. Intelligence Studies*, Vol. 20, No. 2, Fall–Winter 2013.

the development of intelligence program budget submissions, and DNI decision documents, which provide a basis for congressional budget justification books (CBJBs). At the end of the budgeting phase, ODNI staff generate performance and financial results and the NIP portion of the President's Budget, both of which feed into the final phase of the budgeting process. Although the "E" in IPPBE refers to continuous *evaluation,* the final budgeting phase is focused on congressional actions and *execution,* which begin around March of FY(+1) and continue through September of the budgeted fiscal year.[23] The appropriations act is generally signed at the beginning of the fiscal year in October.

Note that Figure 5.3 is a nominal timeline; this is not necessarily how the IPPBE process unfolds in practice. For example, interviewees mentioned that the CIG would often not appear until July, after the MIP budget had already been submitted, creating some challenges that we discuss later in this chapter.[24]

Decisionmakers and Stakeholders

In the following sections, we discuss the evolution of ODNI's IPPBE process from its inception in the 2009–2011 time frame through its current iteration. This discussion refers to decisionmakers and stakeholders across three eras: (1) the Systems and Resource Analysis Office (SRA) era, (2) the Community Resource Investment Board (CRIB) era, and (3) the current Strategic Investment Group (SIG) era. In each of these periods, the overarching themes of evaluation, integration, synchronization, and transparency have remained. However, ODNI staff has adjusted the responsibilities, forums, and approaches involved in executing the IPPBE process.[25]

Two additional factors complicate an analysis of ODNI's IPPBE process in an unclassified report. First, although the process has evolved, the relevant publicly available ICDs have not been updated. This lag time is not unusual, because creating or updating an ICD is a laborious and time-consuming process, involving staff from all components of the IC. Current practice is often captured in briefing documents or other forms of guidance and instruction. We relied on interviews with budget process architects and participants from the three IPPBE eras addressed in this chapter. Second, the IC—by its nature, mission, and authorities—classifies, for counterintelligence reasons, a great deal of information that is readily available for other national and agency budgeting systems. Again, we relied on unclassified interviews with current and former senior intelligence officials and other experts to supplement the available documentation.

---

[23] The role that continuous evaluation plays within IPPBE is to create a structure for periodic links across all other phases. For example, evaluating past strategic investments may result in a DNI decision document that affects the next budget formulation process. Continuous evaluation will be discussed in more detail in the "Budgeting, Execution, and Evaluation" section of this chapter.

[24] ODNI subject-matter experts, interviews with the authors, August 2022–February 2023.

[25] ODNI subject-matter experts, interviews with the authors, August 2022–February 2023.

CFOs, and representatives from the ODNI directorates. The CRIB vetted initiatives and funding requests from across the IC to make recommendations to the PDDNI and the DEXCOM. It was "fed" by an "engine room" staffed by ODNI staff and community members, who analyzed proposals for priority investment.[48] Absent an organization specifically tasked and staffed with a "CAPE-like" capability, the analysis of these proposals was uneven in depth and sophistication.[49] Although the new process increased transparency, it required a tremendous amount of ODNI and IC staff work that was often performed as additional duties.

The intention of the CRIB was to increase transparency and collaboration across the IC elements. This was achieved through greater community involvement and advocacy, with the CFOs involved in the process as facilitators and with decisions made at the DEXCOM level.[50] But, in practice, the CFOs often simply advocated for their respective agencies, and the process often fell short of the goal of generating objective analysis leading to prioritization and decisions that were "best for the community." Furthermore, the engine room proved an imperfect replacement for the previous SRA-driven analytic and evaluation process.[51]

During this period, however, transparency did increase, and the IC adopted a set of integrated priorities identified by the DEXCOM. In some ways, ODNI's IPPBE process moved consciously closer to DoD's PPBE process: The IPG and CIG mirror DoD's defense planning and programming guidance documents, respectively. Similarly, the CRIB and DEXCOM were conceived as functioning in ways analogous to DoD's three-star programmers and DMAG.[52]

Strategic Investment Group Era (2021–Present)

In August 2020, ODNI once again underwent a reorganization, consolidating down to two directorates: Mission Integration (which inherited parts of National Security Partnerships) and Policy and Capabilities (built largely by combining the Strategy and Engagement Directorate and Enterprise Capacity Directorate). This was, in part, a recognition that the staff was simply not resourced to fill out four disparate directorates.[53] The legacy SRA capability, capacity, and missions were absorbed into either the Mission Integration Directorate's Mission Performance, Analysis, and Collection Division or into the Policy and Capability Directorate's Requirements, Cost, and Effectiveness (RC&E) Division. The CFO once again became a direct report to the DNI and PDDNI. Figure 5.4 shows ODNI's updated organizational chart.

---

48 ODNI subject-matter experts, interviews with the authors, October 2022–February 2023.

49 ODNI subject-matter experts, interviews with the authors, October 2022–February 2023.

50 ODNI subject-matter experts, interviews with the authors, October 2022–February 2023.

51 ODNI subject-matter experts, interviews with the authors, October 2022–February 2023.

52 ODNI subject-matter experts, interviews with the authors, October 2022–February 2023.

53 ODNI, "ODNI Implements Organizational Realignment to Increase Effectiveness," press release, August 17, 2020.

FIGURE 5.4
ODNI Organizational Chart as of 2022

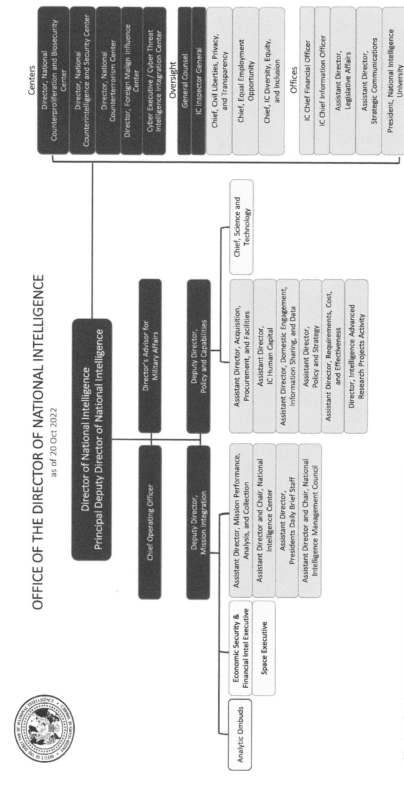

OFFICE OF THE DIRECTOR OF NATIONAL INTELLIGENCE
as of 20 Oct 2022

*This chart reflects the organizational structure as authorized by the Fiscal Year 2022 appropriation.

SOURCE: Reproduced from ODNI, 2022.

With the new organizational structure and the arrival of DNI Avril Haines in January 2021 and PDDNI Stacey Dixon in August of the same year, the IPPBE process continued to evolve to support the priorities of ODNI senior leadership. While the general process and timelines continued to follow ICD 116, there were several modifications in execution based on lessons from the CRIB experience.

For instance, the engine room and CRIB had been conceived as bottom-up approaches, both to optimize transparency and to account for ODNI's inability to maintain a CAPE-like capability. They provided an analytic capability to look across missions, policies, and budgets to make broad recommendations about investment priorities.[54] Unfortunately, the staff had limited time, capacity, and capability to handle the heavy workload, preventing ODNI from providing the kind of deep and thoughtful analysis that would lead the IC to embrace the process.

In contrast to the bottom-up approach, the SIG process is marked by top-down, centralized decisionmaking, driven largely by the DNI, with decisions made at her level in consultation with the SIG members.[55] The SIG itself consists of relevant heads of the IC agencies or their representatives, is chaired by the DNI, and is staffed by the ODNI's Deputy Director for Policy and Capabilities. The SIG process relies on "expert groups" drawn from the IC agencies to provide analysis to the SIG members. Such a process can lead to the perception that the agencies are "grading their own work," because the Executive Committee (EXCOM)—a senior advisory group led by the DNI and consisting of the IC element directors—and DEXCOM can act only on information provided to them by the agencies.[56] Compared with the CRIB process, the SIG process reduces the role of the IC CFOs. The reduced number of people at the decisionmaking table, compounded by the largely virtual format because of the COVID-19 pandemic, has resulted in a process that limits transparency below the most senior level.[57] Currently, budgetary decisions are made by only a small number of individuals at the EXCOM and DEXCOM levels.[58] However, the advantage of this system is that the part of the IC budget reserved for community investment can be rapidly and effectively focused on a set of "game-changing" community capabilities—agreed on at the most senior level of the IC—something that would not be possible for the agencies otherwise.

While these are important differences between the SIG era and its predecessor SRA and CRIB eras, the broad outlines of ICD 116 remain, with responsibilities shifted and assumed by new organizations in the evolved ODNI organizational chart. The framers of ICD 116

---

54 ODNI subject-matter experts, interviews with the authors, October 2022–February 2023.

55 ODNI subject-matter experts, interviews with the authors, October 2022–February 2023.

56 The EXCOM advises and supports the DNI, conducts in-depth discussions on critical issues, and enables resource allocation. Prior to the creation of ODNI, no such IC-wide senior leadership advisory forum existed. See ODNI, "ODNI Fact Sheet," October 2011c.

57 ODNI subject-matter experts, interviews with the authors, October–December 2022.

58 ODNI subject-matter experts, interviews with the authors, August 2022–November 2022.

wanted to use the IPPBE process to demonstrate the DNI's role and authorities. If anything, the SIG process probably demonstrates that authority more than even the SRA process did, largely as a function of DNI focus and interest.

The ODNI staff continues to focus on various forms of evaluation, many of which remain or are analogous to the responsibilities and products described in ICD 116. The staff is more synchronized around the budget cycle and the necessary budget inputs from the two directorates and the CFO staff than it was in the SRA era, when many functions were performed only by SRA without involvement by the broader staff, or when the Intelligence Integration Directorate and NIMs were sometimes working at cross-purposes. The prioritizing and integrating activities remain important parts of the IPPBE process.

Next, we go into more detail on specific aspects of ODNI's planning, programming, and budgeting process, paying particular attention to the outputs of each phase, information flows, timelines, and the level at which budgetary decisions are made.

Planning and Programming

ODNI's planning process begins with the identification of strategic priorities and MILs, which ultimately are winnowed down to the MISs. These studies systematically analyze the costs, performance, and benefits of alternative investments. Together, the strategic priorities and systematic analyses inform the DNI's resource decisions.[59]

In the planning phase, ODNI seeks to chart the strategic direction for both the NIP and MIP. With the codification of the IPPBE process in ICD 116 in 2011, resource decisionmaking was moved up and to the left—meaning that the planning process began earlier, and decisions were made at higher echelons. Specifically, the planning process was shifted to the left by almost a year.[60] The top row of Figure 5.3 depicts how both the planning and programming phases begin in FY(+3), three years before the year of execution.

The SRA process generated mixed reviews when it came to planning and programming. Some IC elements believed that the SRA construct gave the IC agencies a greater ability to understand their budgetary requirements and resource allocation needs.[61] However, other IC elements felt that the SRA construct encroached on their programming involvement.[62]

The CRIB process increased IC element CFOs' involvement in planning and programming, resulting in an increased sense of transparency.[63] In essence, the IC element CFOs pushed information up to ODNI and provided feedback in a way that mirrored the program

---

[59] ODNI, "IC Planning Process," undated-b.

[60] ODNI subject-matter expert, interview with the authors, August 2022.

[61] ODNI subject-matter experts, interviews with the authors, October and December 2022.

[62] ODNI subject-matter expert, interviews with the authors, October 2022.

[63] ODNI subject-matter experts, interviews with the authors, November 2022.

and budget review process in the OSD, which provides top-line guidance for the programming phase of DoD's PPBE process.[64]

The SIG process limits the involvement of IC element CFOs; their roles are now largely confined to facilitating resource information-sharing to ODNI to support the development of the CIG.[65] It is unclear how ODNI uses the resource information inputs from the IC elements, as decisionmaking has been confined to a smaller number of high-level officials.[66]

Key products produced in the planning and programming phases include the following:[67]

- National Intelligence Strategy (P&S)
- IPG (Mission Integration Directorate)
- Defense Intelligence Strategy (OSD)
- strategic assessments (P&S)
- Annual Threat Assessment of the U.S. Intelligence Community (ODNI)
- SERs (RC&E)
- functional managers' annual reports (IC elements)
- MISs (RC&E)[68]
- National Intelligence Strategy progress assessment (P&S)
- NIP summary of performance and financial information (ODNI CFO)
- various enterprise-wide assessments and DNI-directed studies (ODNI staff).[69]

The IC planning phase consists of five steps:

1. *Strategic analysis.* In step 1, the MIL is developed in May of each year, and the DNI determines which MISs are included in the draft CIG. In alternating years, the Mission Integration Directorate engages the NIMs and the ODNI staff to prepare the IPG, which looks across regions and functions to determine priorities, gaps, and shortfalls in analysis and collection; areas of acceptable risk; and potential mitigations.

2. *Prioritization and capabilities-based analysis.* Step 2 involves collecting data on future capability needs identified by the NIMs. These needs are consolidated and aligned

---

[64] Defense Contract Management Agency Manual 4503-02, *Programming,* U.S. Department of Defense, February 21, 2020, p. 10; ODNI subject-matter expert, interview with the authors, November 2022.

[65] ODNI subject-matter experts, interviews with the authors, November 2022.

[66] ODNI subject-matter experts, interviews with the authors, November and December 2022.

[67] ODNI, undated-b. The offices, agencies, and organizations shown in parentheses are those with responsibility for the specified products.

[68] MISs were originally performed by SRA and may still be done by RC&E or another ODNI organization, as applicable, depending on the subject. However, other products to support decisionmaking were produced by the engine room for the CRIB, now the expert groups for the SIG.

[69] This is subject- and purpose-dependent and could be produced by RC&E; Acquisition, Procurement, and Facilities; or Mission Performance, Analysis, and Collection.

with the National Intelligence Strategy and reflected in the IPG. The consolidated priorities feed into both the requirements development process and the development of future MISs. ODNI informs IC elements of areas identified for disinvestment while respecting each program's responsibility to manage its activities.

3. *Integration.* In step 3, the priorities identified in step 2 are integrated across the NIP and MIP. The NIP-MIP Integration Group leads this integration effort and deconflicts proposed strategic priorities.

4. *Approval.* Step 4 involves obtaining OSD's concurrence for the MIP items. Then, the DNI finalizes the strategic priorities and approves the IPG.[70]

5. *Finalization.* Step 5 culminates with the draft CIG document, which directly feeds into the programming phase of the IPPBE process.[71]

Generally, the planning phase provides boundaries and guidance for the programming phase. The intent of the programming phase is to provide the DNI with resource allocation options that adhere to the final CIG. The planning and programming phases have significant overlap.[72] As shown in Figure 5.3, the planning phase begins around March of FY(+3) and ends around December of FY(+2); the programming phase begins around May of FY(+3) and ends around May of FY(+2).[73] The role of P&S in the process overlaps with the programming and budgeting phases to ensure that current IC activities align with the DNI's policy, strategy, and vision.[74]

Budgeting, Execution, and Evaluation

IRTPA assigns budgetary authority for the NIP to ODNI and the DNI.[75] The ODNI CFO supports the DNI in formulating and justifying the NIP, overseeing budget execution, developing budget performance objectives, promoting IC fiscal accountability, and improving finan-

---

[70] The IPG is published every other year and may be updated in off-years as necessary (ODNI subject-matter expert, interview with the authors, February 2023).

[71] ODNI, undated-b; ODNI subject-matter experts, interviews with the authors, November 2022–February 2023.

[72] David Luckey, David Stebbins, Sarah W. Denton, Elena Wicker, Stephanie Anne Pillion, and Alice Shih, *National Geospatial-Intelligence Agency Resources: Financial Management Programming Evaluation*, RAND Corporation, RR-A659-1, 2022. Although this resource is focused on the National Geospatial-Intelligence Agency's programming phase, challenges and general findings related to the overlap between planning and programming phases apply across the IC enterprise.

[73] ODNI, undated-c.

[74] ODNI, "Policy & Capabilities–Who We Are," webpage, undated-e.

[75] Pub. L. 108-458, 2004.

cial management.[76] The specific responsibilities of the ODNI CFO include the following:[77]

- drafting the performance, procedural, and fiscal guidance for the NIP
- advising the DNI and senior leadership on budget formulation, justification, and execution
- building and defending the NIP through DNI decision documents, CBJBs, budget testimony, appeals, supplemental appropriations, and OMB top-line requests
- leading IC intelligence program budget submission reviews
- managing budget performance planning through the development of an annual IC report, quarterly scorecard, metrics, and performance reviews
- overseeing the execution of the NIP budget through apportionment documents, reprogramming actions, and quarterly congressional budget execution reports
- participating in the development of the MIP
- leading IC financial auditing efforts and coordinating agency financial reports
- developing IC financial management policy guidelines and standards
- liaising with the IC, OMB, DoD, and Congress on budgetary matters.

The budgeting phase of the IPPBE process begins with the draft CIG document, which is usually provided to the IC elements in February or March. Around May, the final CIG from the programming phase directly informs procedural guidance, which is the first output of the budgeting phase. This guidance is aligned with the MISs, integrated with intelligence program budget submissions, and provided to the DNI for adjustments through DNI decision documents.[78] The CFO leads the development of the CBJB for OMB review. The budgeting phase ends with the production of the performance and financial results and the NIP portions of the intelligence budget.[79]

The ODNI CFO also monitors the execution of IPPBE, which includes congressional actions, appropriations, reprogramming, and supplemental appropriation requests.[80] However, execution responsibility ultimately rests with the IC elements, which allocate the NIP appropriations provided to them by the DNI.[81] The ODNI budget is typically appropriated annually (i.e., one-year money).

---

[76] ODNI, "Chief Financial Officer," webpage, undated-a.

[77] ODNI, undated-a.

[78] ODNI subject-matter experts, interviews with the authors, November 2022–February 2023.

[79] ODNI subject-matter experts, interviews with the authors, November 2022–February 2023.

[80] Like with other organizations, including DoD, continuing resolutions affect ODNI's ability to start new work and projects.

[81] ODNI subject-matter expert, interview with the authors, February 2023.

ODNI's appropriations structure mirrors that of DoD, including its appropriation account categories.[82] While there have been discussions about adopting something like the common appropriations structure that has allowed DHS to consolidate its accounts into four categories, ODNI's defense appropriations preclude the NIP's consolidation in annual appropriations bills.[83] Moreover, migration to such a consolidated structure would have significant counterintelligence implications.[84] Some observers have hypothesized that two-year appropriations may increase ODNI budgetary flexibility, but such a construct would not necessarily provide equal benefit for IC elements that are not "operation and maintenance–heavy" and more focused on acquisitions; other observers contend that annual appropriations afford better fiscal control.[85]

To monitor budget inputs and analysis, the ODNI CFO uses IRIS, a data management system.[86] At the time of this research, the next generation of IRIS was being built and tested.[87] Yet, each IC element has its own budget monitoring system, which must be integrated into IRIS, and each element provides only budgetary estimates (i.e., not direct links to budget items).[88]

Unique to IPPBE, the execution phase was intentionally designed to flow into a continuous evaluation process. The goal of continuous evaluation is to create a positive feedback loop that links budget execution to the next planning process. P&S's role in continuous evaluation is to assess progress toward achieving the DNI's vision as set forth in the National Intelligence Strategy. During the SRA era, in particular, SERs were intended to look backward to assess the effectiveness of resource allocations.[89] Although SERs do provide the opportunity for the DNI to incorporate findings from these evaluations into DNI decision documents, they can take two years to complete, reducing the time between such a strategic evaluation of resources and the next budgeting cycle.[90] The CFO's budget performance reports provide an additional means of evaluating ODNI's resource allocations. The dark gray boxes at the

---

[82] We are unable to discuss appropriation account apportionments with any specificity or detail. Such information is not available to the general public and has counterintelligence implications.

[83] ODNI subject-matter expert, follow-up discussion with the authors, February 17, 2023; Painter, 2019.

[84] ODNI subject-matter expert, follow-up discussion with the authors, February 2023. Further detail on this topic is not available to the general public.

[85] ODNI subject-matter experts, interviews with the authors, October 2022–February 2023.

[86] ODNI subject-matter expert, interview with the authors, February 2023.

[87] ODNI subject-matter expert, interview with the authors, February 2023.

[88] ODNI subject-matter expert, interview with the authors, February 2023.

[89] ODNI subject-matter expert, interview with the authors, September 2022.

[90] Another challenge mentioned by interviewees is that SERs required analytic capability and expertise that was not available within ODNI, and the findings of SERs could have negative implications for the IC broadly (ODNI subject-matter experts, interviews with the authors, August–October 2022).

bottom of Figure 5.3 contain the various inputs and roles in the continuous evaluation process that make IPPBE distinct from PPBE.

Oversight

The ODNI CFO manages engagement with oversight bodies, such as Congress and OMB.[91] Intelligence Community Standard (ICS) 104-01 stipulates that, in accordance with OMB Circular No. A-11, IC element CFOs must seek guidance from the ODNI CFO when discussing the information in the CBJB.[92] Although the IC element CFOs have a responsibility to keep congressional committees informed of NIP and financial management operations, they also have a responsibility to coordinate with the ODNI CFO prior to any engagement with Congress.[93] This process ensures that the IC speaks to OMB and Congress with a single voice.[94]

At the end of the budgeting phase, the ODNI CFO coordinates with OMB and congressional appropriations committees on the CBJB. *OMB passback* (see Figure 5.3) refers to a prioritization process between OMB and the ODNI CFO in which OMB's focus is on ensuring that the President's budgetary priorities are addressed—not on ODNI or IC element missions. Given that OMB's authorities stem from the Budget and Accounting Act of 1921 and Article 2 of the U.S. Constitution, OMB footnotes and apportionments—that is, OMB's comments and inputs on the CBJB—can generate some hostility between OMB and ODNI.[95] Once OMB and ODNI reach agreement on the CBJB, it is presented to Congress for appropriation actions as a consolidated NIP budget.

IRTPA provided Congress with a means of centralizing oversight of this process to some extent—not dissimilar from the way the National Security Act of 1947 and the Goldwater-Nichols Act of 1986 gave the House and Senate Armed Services Committees some oversight of the Secretary of Defense and Chairman of the Joint Chiefs of Staff. Since IRTPA's passage, Congress can use the DNI to address issues that span multiple IC elements. Rather than engaging multiple IC elements on a single issue, IRTPA has allowed Congress to rely on ODNI for data collection and analysis that can inform congressional decisionmaking more broadly.[96]

---

[91] ICD 104, 2013.

[92] ICS 104-01, *Engagement with Congress on National Intelligence Program (NIP) Budgeting Activities*, Office of the Director of National Intelligence, June 27, 2018.

[93] ICS 104-01, 2018.

[94] ICS 104-01, 2018.

[95] ODNI subject-matter expert, interview with the authors, February 2023. Also see U.S. General Accounting Office, *The Budget and Accounting Act*, 1966; and U.S. Constitution, Article II, Section 3.37, Impounding Appropriated Funds.

[96] Rosenwasser, 2021. These perspectives were also shared by an ODNI subject-matter expert in an interview with the authors in February 2023.

There are two congressional committees with primary oversight jurisdiction for the NIP: the Senate Select Committee on Intelligence and the House Permanent Select Committee on Intelligence (HPSCI).[97] For the HPSCI, budget monitors are designated from the majority party to directly engage with the ODNI and the IC elements on their appropriations:

> IRTPA anticipates guidance from ODNI to the heads of [IC elements] during preparation of the NIP based on White House priorities. Congress appropriates funds to the various NIP agencies accordingly.[98]

Regional and functional monitors look across the IC to ensure integrated oversight.[99] Congressional budgetary actions can fence, add, or remove funding or positions from a program.[100] Close working relationships between ODNI, the IC elements, and congressional committees of jurisdiction can promote a positive relationship between the DNI and Congress. However, the relationship between ODNI and Congress varies greatly from committee to committee, topic to topic, and member/monitor to member/monitor.[101]

With respect to reprogramming funds, ICS 104-02 cites the DNI's responsibility for managing the NIP in accordance with Title 50 and outlines the process for adjusting NIP allocations during the execution year.[102] Reprogramming begins with the CFO establishing baseline figures by which adjustments to NIP resources are tracked. These adjustments occur at the NIP project level and are reported to Congress after consultation with OMB and IC elements. ICS 104-02 states that NIP funds may be reprogrammed when

- funds are transferred to a high-priority intelligence activity in support of an emergent need
- funds are not moved to a reserve for contingencies of the DNI or the CIA
- the reprogramming results are cumulatively less than $150 million and less than 5 percent of accounts available to a department or agency under the NIP in a single fiscal year[103]
- the action does not terminate an acquisition program
- the congressional notification period is satisfied.[104]

---

[97] ODNI subject-matter expert, interview with the authors, February 2023. The Senate Committee on Armed Services and the House Committee on Armed Services have oversight responsibility for the MIP.

[98] Luckey et al., 2022, p. 24.

[99] ODNI subject-matter expert, interview with the authors, February 2023.

[100] ODNI subject-matter expert, interview with the authors, February 2023.

[101] ODNI subject-matter expert, interview with the authors, February 2023.

[102] ICS 104-02, *National Intelligence Program (NIP) Procedures for Reprogramming and Transfer Actions*, Office of the Director of National Intelligence, June 26, 2018.

[103] ICS 104-02 states that a reprogramming action can exceed these amounts if the department head or CIA director concurs with the action.

[104] ICS 104-02, 2018.

TABLE 5.1

Summary of Lessons from ODNI's Budgeting Process

Theme	Lesson Learned	Description
Planning and programming	Lesson 1: ODNI processes facilitate prioritization and risk mitigation.	The IPPBE process has used various means to address emergent strategic priorities and provide a means for ensuring that risk management is sufficient and integrated into resource decisionmaking.
Budgeting, execution, and evaluation	Lesson 2: Continuous evaluation is an integral part of ODNI's budgeting process.	Backward-facing and forward-looking continuous evaluation is critical to enhancing the effectiveness of resource allocation decisions.
Oversight	Lesson 3: ODNI engages senior leaders in all budgetary decisions.	Decisionmaking at higher echelons helps ensure that the IPPBE process is integrated across the complex IC enterprise.

CHAPTER 6

Key Insights from Selected Non-DoD Federal Agencies Case Studies

The four case studies presented in this report provide the Commission on PPBE Reform with insights into how other non-DoD U.S. federal government agencies navigate U.S. political institutions and resource planning processes to meet mission needs. In Chapters 2 through 5, we discussed how these agencies conduct defense resource planning, programming, budgeting, execution, and oversight—and the strengths and challenges of their approaches.

This final chapter focuses on summary takeaways. As part of this analysis, we used an initial set of standard questions from the commission, focusing on core areas related to resource planning, as a means of ensuring that there would be some ability to compare across cases. The material presented in this chapter, distilled from Chapters 2 through 5, outlines important themes for the commission to understand when trying to compare DoD's defense resource planning processes with those of other U.S. government agencies. Despite significant differences between DoD and these selected agencies in terms of mission requirements, portfolio, organizational evolution, oversight, and size, among many other factors, these cases suggest several insights that are germane for DoD, which we present below.

The following section on key insights consolidates the strengths, challenges, and lessons outlined in the case studies in this volume. The concluding section on applicability speaks directly to the commission's mandate—and to the potential utility of these insights for DoD's PPBE System.

Key Insights

Key Insight 1: Other U.S. Government Agencies Looked to DoD's PPBE System as a Model in Developing Their Own Systems, Which Subsequently Evolved

In 1965, President Johnson drew on the still-nascent PPBS in DoD as a model for the implementation of analogous systems across the federal government. Although that formal effort fizzled out a few years later, all the agencies considered in this report, as the names of their budget processes often suggest—NASA's PPBE, ODNI's IPPBE, DHS's PPBE, and HHS's budget process—have looked to DoD's PPBE System as a model for a structured and mature approach to planning and resource allocation decisionmaking.

However, budget processes have evolved individually in accordance with the agencies' missions, organizational structures, authorities provided by Congress, staff capacities, other available resources, and many other factors. For example, although some vestiges of the PPBE framework—such as its rigorous program evaluation capabilities—remain features of the contemporary HHS budgeting system, the department gradually dismantled much of its PPBS during the 1970s in response to the perception that PPBS did not fit with HEW's missions, organizational structure, or program needs.[1] HHS's budgeting system has, therefore, diverged significantly from DoD's since 1980.

ODNI's IPPBE process was specifically modeled on DoD's PPBE process, with some modifications to ensure that it was suited to ODNI's mission. This adaptation was done both by necessity, to account for the presence of DoD elements in the IC, and by design, given that the architects of the IPPBE process were well versed in PPBE and viewed it as the best model for ODNI's complex organizational structure.[2] One notable and deliberate difference between the IPPBE and PPBE processes is ODNI's substitution of *evaluation* for DoD's *execution*.[3]

Despite the evolution away from DoD's PPBE framework, all four agencies still generally follow a budgeting process that is common to most U.S. federal civilian agencies. This process begins with an annual planning cycle and culminates in budget execution and performance evaluation.

Key Insight 2: Long-Term Planning Is Often Limited Relative to That Done by DoD

One difference between DoD and three of the agencies considered in this report is DoD's focus on long-term planning processes. We attribute this difference both to the inherently dynamic requirements of DHS's and HHS's mission sets and to the weaker (relative to DoD) mechanisms for forging forward-looking, cross-departmental plans through a headquarters function in DHS and ODNI. (NASA, in contrast, adheres to five-year planning guidance and decadal studies to identify future requirements).

Because HHS programs deliver mostly health care services and grants, outside its mandatory funding, the department operates primarily on one-year discretionary funding and restricts budget planning to the annual budget cycle.[4] Consequently, HHS does not focus on long-term budget planning, nor does it have well-established links between strategic plan-

---

[1] See, for examples, Harlow, 1973, p. 90; Jablonsky and Dirsmith, 1978, p. 216; Rivlin, 1969, p. 922; and U.S. General Accounting Office, 1990, p. 22.

[2] ODNI subject-matter experts, interviews with the authors, August–September 2022.

[3] ODNI subject-matter experts, interviews with the authors, August–September 2022.

[4] HHS officials, interviews with the authors, October 2022–January 2023.

ning and budgeting.[5] Long-term planning is particularly important for agencies with missions requiring sustained development efforts rather than short-term operational programs.

DHS's federated model, in which the components remain responsible for their separate missions and receive direct appropriations, makes it difficult to promote cross-departmental priorities, unlike in DoD, where OSD disperses the funds, increasing its control. DHS also lacks a Goldwater-Nichols initiative to compel jointness, and DHS headquarters lacks the resourcing of OSD, limiting the ability of DHS headquarters to coordinate and manage its components.

Key Insight 3: A Variety of Mechanisms Enable Budget Flexibility and Agility

Mechanisms have been designed to meet dynamic mission demands, such as the startup of DHS; provide flexible authorities to meet highly variable mission needs, as with DHS's Disaster Relief Fund and CDC's Infectious Disease Rapid Response Reserve Fund; and adapt quickly to emerging public health threats, as with HHS's Public Health Service Evaluation Set-Aside.

We identified several other mechanisms for enabling flexibility and agility, primarily by giving agencies more discretion (than in DoD) to redirect appropriated funds. HHS and NASA benefit from having fewer restrictions on fungible expenditures, which allows them to shift resources in accordance with changing priorities. The LHHS appropriations bills afford HHS considerable flexibility in reprogramming funds: Below a reprogramming threshold of the lesser of $1 million or 10 percent of a budget account, HHS is not required to report it to Congress.[6] Congress has granted HHS additional sources of flexibility during budget execution, including the Secretary's One-Percent Transfer General Provision, which allows HHS to transfer up to 1 percent from any LHHS appropriation account into another account up to 3 percent of the amount of the receiving account, with a maximum transfer amount of around $900 million.[7] HHS appears to have wide latitude in how appropriated funds are spent. Similarly, NASA does not appear to receive appropriations in distinct titles, as does DoD. In part, this is likely due to NASA's narrower mission requirements (e.g., NASA does not procure at the scale of DoD).

Another mechanism for flexibility is the authority to carry over funding across years. Although DHS's no-year appropriations have been curtailed, DHS still has authority to allow its components to carry over and spend in the next fiscal year up to 50 percent of prior-year balances in one-year O&S accounts. HHS also has authority to repurpose expiring unobligated balances. The NEF allows HHS to take expired, unobligated funds and reallocate them

---

[5] HHS officials, interviews with the authors, October 2022–January 2023.

[6] NIH, Office of Management and Assessment, 2020.

[7] The Office of Refugee Resettlement can take up to 15 percent of the value of transfer funds, so these funds are often transferred there (HHS official, interview with the authors, January 2023).

to a department-wide capital investment account. HHS has used the NEF extensively to fund IT systems, particularly for cybersecurity purposes, but operating and staff divisions can request funding for other capital expenditures as well.

In some instances, Congress further enables agility by employing broader appropriation categories than those used for DoD appropriations; in this way, agency decisionmakers have more flexibility to implement changes to previously communicated funding priorities. For instance, NASA's receipt of appropriations primarily at the mission level allows mission directorates to decide how to allocate funding between projects without having to seek additional congressional approval.

Key Insight 4: Mechanisms for Enabling Agility Help Agencies Weather Continuing Resolutions and Other Sources of Budget Turbulence

Just as budget flexibilities, such as those cited above, can let a manager decide how to prioritize and where to take risks in light of changing mission needs, they can also help an agency manage under continuing resolutions. NASA's two-year expiration timeline for appropriations reportedly provides the agency with a cushion in the likely event that a regular appropriation is delayed.

Continuing resolutions across the U.S. government remain painful, but an agency's ability to manage them is a function of its portfolio, investments, actions, and other factors. For example, HHS develops requests for grant proposals ahead of anticipated continuing resolutions.[8] The ability of DHS components to carry over into the next fiscal year one-year O&S accounts and expend up to 50 percent of their prior-year lapsed balances could also help DHS mitigate the effects of continuing resolutions, although, as mentioned previously, this is not the main purpose of carryover funding.

A different kind of example for HHS is that its mandatory funding (primarily for Medicare and Medicaid) constitutes about 90 percent of the HHS budget.[9] Most mandatory HHS programs, such as Medicare and children's entitlement programs, are budgeted on ten-year schedules outside the annual appropriations process and, thus, are rarely subject to continuing resolutions. Medicaid, however, is still subject to annual congressional oversight.[10]

---

[8] GAO, 2021.

[9] Tollestrup, Lynch, and Cornell, 2022, p. 4.

[10] Tollestrup, Lynch, and Cornell, 2022, p. 4.

Key Insight 5: The Latest Replacement of Execution with Evaluation in PPBE-Like Processes Could Be Instructive for DoD

ODNI is not alone in substituting *evaluation* for *execution* in its budgeting process. DHS has also essentially done so in its PPBE-like process to better understand the results of its spending.

To support implementation of the 2018 Evidence Act, DHS issued a policy on program and organizational evaluations in February 2021 and annual evaluation plans for FYs 2022 and 2023. This line of effort demonstrates an investment by DHS in evaluation activities. DHS's efforts in this area could help inform DoD's approach to the execution phase.

Key Insight 6: Implementation of PPBE-Like Processes at the Scale of DoD's Process Is Resource-Intensive, Institutionally Challenging, and Often Infeasible for Smaller Agencies

One area in which the selected non-DoD agencies cannot emulate an exemplary DoD PPBE capability is DoD's CAPE analytic function. In these four agencies, a CAPE-like function does not exist in comparable size and mission, because this function is resource-intensive to build and maintain and challenging to empower institutionally. CAPE's mission is to "provide the Department of Defense with timely, insightful and unbiased analysis on resource allocation and cost estimation problems to deliver the optimum portfolio of military capabilities through efficient and effective use of each taxpayer dollar."[11]

By comparison, the planning, programming, and budgeting for NASA are handled by one NASA organization—the OCFO—and there is a possibility that conflicts of interest might arise. The OCFO's Strategic Investment Division, which develops strategic planning guidance that informs NASA's programming and budgeting phases, is not analogous to CAPE because it is housed within OCFO and, therefore, may not be considered an independent organization when it scrutinizes NASA's budget submissions. Interviewees noted that independent agencies may be able to identify overruns faster than NASA's evaluation-oriented offices could, primarily because of workforce capacity challenges.[12]

ODNI attempted to emulate the analytic rigor of the CAPE function by creating SRA and tasking it with managing IPPBE integration and synchronization. The goal was to establish a predictable, transparent, and repeatable method to collect and prioritize critical intelligence requirements and to translate those priorities into resource allocations through the IPPBE process. Between 2009 and 2011, ODNI tested this process, which was later codified. But several years later, concerns arose over the evolution of some documents and the ability of SRA to continue its role. ODNI found it difficult to keep SRA staffed with the capability and capacity to fulfill a CAPE-like function.

---

[11] OSD, Cost Assessment and Program Evaluation, undated.

[12] NASA subject-matter experts, interviews with the authors, September 2022.

Key Insight 7: Consolidated Resource Management Information Systems Could Improve Visibility Across the Federated Structures of Government Agencies

DHS's consolidation of its PPBE information system has enhanced its ability to create and manage budgets. DHS officials reported that the consolidated system for generating congressional budget justification documents, developing a five-year funding plan, and capturing performance management data has reduced their reliance on Microsoft Excel spreadsheet templates and data reentry, allowing DHS to automate the generation of certain reports that were previously created manually.

Although HHS does not have a consolidated resource management information system, some HHS operating divisions have constructed their own. Still, at the department level, the lack of a consolidated budget formulation system has left HHS leadership with limited visibility into OPDIVs' budgets. Because DoD is a comparably federated organization with diffuse authorities, it could explore the feasibility, costs, and benefits of constructing a consolidated PPBE information system and whether the benefits of doing so would outweigh the costs.

Applicability of Key Insights to DoD's PPBE System

The Commission on PPBE Reform is looking for potential lessons from the PPBE-like systems of non-DoD federal agencies. While those agencies' budgeting processes were originally modeled after DoD's PPBS, they have adapted their processes to align with the unique missions of each agency. Despite the movement away from DoD's PPBS model, the agencies still use similar PPBE processes. Because of these similarities, there would be no benefit from DoD adopting any of these systems wholesale. However, there is value in exploring the ways in which Congress provides each agency with flexibility so that DoD can ask for similar kinds of flexibility to support more innovation, make funding more predictable over multiple years, and obtain relief from various pain points in the system. These pain points include continuing resolutions, rigid appropriations categories, and appropriations for line items instead of portfolios. The commission could further explore the flexibility mechanisms identified below, organized by agency.

DHS funds are typically budgeted annually, but some programs receive multiyear or no-year appropriations. Congress sometimes appropriates multiyear funds to major acquisition programs to foster a stable production and contracting environment. A key example of no-year money is the Disaster Relief Fund, which is meant to give FEMA the flexibility to respond quickly to emerging disaster relief and recovery needs. As another example, DHS officials mentioned how the border security, fencing, infrastructure, and technology appropriation gave DHS the ability to carry over significant amounts of funds related to this mission area. (DHS officials noted that funds are no longer appropriated to this account and that the use of no-year appropriations was significantly curtailed with the implementation of the common appropriations structure.)[13] Congress also authorizes DHS components to carry one-year O&S accounts forward into the next fiscal year and to expend up to 50 percent of the prior-year lapsed balance amounts. Beyond the base budget, DHS often receives supplemental funds for emergent requirements, the number of which varies from year to year.

HHS has access to emergency supplemental funding and several flexible-spending accounts, such as the NEF, which allows HHS to reallocate expired, unobligated funds to capital investments. These flexibility mechanisms are often given multiyear or no-year funding. HHS does not use a common appropriations structure, so budget justifications focus heavily on missions and needs. This focus allows discussions between the OPDIVs and the SBC's department-level leadership to concentrate on aligning program budgets and missions with the HHS Secretary's priorities.

NASA requests and is allocated funding differently than DoD. Because NASA's funds are appropriated to mission directorates primarily at the mission, theme, and project levels, NASA has some flexibility to align project funding to meet changing priorities or real-world circumstances. NASA's FY 2023 congressional justification does not request, nor is it funded with, appropriations split into categories, such as RDT&E, procurement, and O&M in the same manner as DoD. Therefore, NASA does not appear to encounter the same types of restrictions as DoD with respect to using specific funding for specific activities (e.g., using RDT&E only during the design and development stages of a program). Moreover, all of NASA's appropriations, except for construction, have two-year durations. NASA has obligation goals of 90–95 percent in the first year of two-year funds, which allow for some funding to be expended in the second year, typically at the start of the fiscal year. Because continuing resolutions are a real possibility, this carryover funding can mitigate any shortfalls that might result at the start of a fiscal year—and, thus, act as a cushion for continuing resolutions.

ODNI funds may be reprogrammed under five conditions: (1) when funds are transferred to a high-priority intelligence activity in support of an emergent need, (2) when funds are not moved to a reserve for contingencies of the DNI or the CIA, (3) when funds are cumulatively less than $150 million and less than 5 percent of the annual accounts available to a department or agency, (4) when the action does not terminate an acquisition program, and (5) when the congressional notification period is satisfied. Congress must be notified of above-the-threshold reprogramming actions (i.e., those that exceed $150 million or 5 percent) within 30 days or within 15 days for matters of urgent national security concern. Below-the-threshold reprogramming actions do not require congressional notification. However, ODNI does notify Congress of below-the-threshold actions that may be of congressional interest.[14]

---

[13] DHS officials, interview with the authors, November 2022. For more information on the border security, fencing, infrastructure, and technology appropriation, see Painter and Singer, 2020.

[14] ICS 104-02, 2018.

Summary of the Budgetary Flexibilities of Comparative U.S. Federal Agencies

In Tables 6.1 through 6.4, we summarize the budgetary flexibilities of the assessed non-DoD U.S. federal agencies, compared with DoD budgetary flexibilities.[15] As an introduction, Table 6.1 specifies each agency's planning and budget system. Table 6.2 summarizes the funding categories and funding availability within each system. Table 6.3 compares the different types of carryover funds and restrictions during continuing resolutions. Table 6.4 focuses on the different kinds of reprogramming, transfers, and supplemental funding available within each system.

TABLE 6.1

Planning and Budget Systems of DoD and Comparative U.S. Agencies

Agency	Planning and Budget System
DoD	Planning, Programming, Budgeting, and Execution (PPBE) System
DHS	Future Years Homeland Security Program (FYHSP)
HHS	No direct analog at departmental level; operating divisions (OPDIVs) have individual approaches to annual budget planning and formulation
NASA	PPBE System
ODNI	Intelligence Planning, Programming, Budgeting, and Evaluation (IPPBE) System

TABLE 6.2

Funding Categories and Funding Availability for DoD and Comparative U.S. Agencies

Agency	Funding Categories	Funding Availability
DoD	• Discretionary budget includes Military Personnel (MILPERS), Operation and Maintenance (O&M), Procurement, Research, Development, Test, and Evaluation (RDT&E), and Military Construction (MILCON) account categories	• Varies by account type; multiyear or no-year appropriations for limited programs as authorized by Congress
DHS	• Discretionary budget includes component-level accounts organized by four common categories • Mandatory funding for some functions, such as Coast Guard benefits • Some activities funded through discretionary fees and collections	• Varies by account type; multiyear or no-year appropriations for certain programs as authorized

---

[15] Information presented in these tables is derived from multiple sources and materials reviewed by the authors and cited elsewhere in this report. See the references list for full bibliographic details.

Table 6.2—Continued

Agency	Funding Categories	Funding Availability
HHS	• Discretionary budget organized under 12 OPDIVs • Mandatory funding is ~90% of budget • Some activities funded through discretionary fees	• One-year appropriations for most of discretionary operational budget; multiyear and no-year appropriations for certain programs
NASA	• Discretionary budget with output-oriented appropriations allocated at program level	• Six-year appropriations, construction • Two-year appropriations (except Office of Inspector General and Construction and Environmental Compliance and Restoration), all other account types
ODNI	• Discretionary budget for National Intelligence Program (NIP) activities managed by ODNI • Discretionary budget for Military Intelligence Program (MIP) activities managed through DoD	• Varies by account type; one-year appropriations for ODNI operations

TABLE 6.3

Carryover Funds and Restrictions for DoD and Comparative U.S. Agencies

Agency	Carryover Funds	Restrictions During Continuing Resolutions
DoD	Limited carryover authority in accordance with Office of Management and Budget Circular A-11	Various; no new programs, increases in production rates, etc.
DHS	Authority to carry over one-year operations and support funding into the next fiscal year; can expend up to 50% of prior-year lapsed balance	Various; no new programs, new hiring, or new contract awards for discretionary programs.
HHS	Limited carryover authority in accordance with OMB Circular A-11	Various; new contract awards and grants have been suspended for discretionary programs.
NASA	Limited carryover authority in accordance with OMB Circular A-11	Minimal; two-year appropriations and 90–95% obligation goal for first year of availability allow forward funding of contracts.
ODNI	Limited carryover authority in accordance with OMB Circular A-11	Restrictions on ODNI/NIP operations are unclear; MIP operations are subject to restrictions on DoD activities during continuing resolutions.

TABLE 6.4

Reprogramming, Transfers, and Supplemental Funding for DoD and Comparative U.S. Agencies

Agency	Reprogramming	Transfers	Supplemental Funding
DoD	• As authorized; four defined categories of reprogramming actions • Prior-approval reprogramming actions—increasing procurement quantity of a major end item, establishing a new program, etc.— require approval from congressional defense committees	As authorized; general and special transfer authorities, typically provided in defense authorization and appropriations acts	Frequent; linked to emerging operational and national security needs
DHS	• As authorized; request to Congress must be made before June 30 if additional support for emerging needs or crises exceeds 10% of original appropriated funding • Restrictions (creation of program, augmentation of funding in excess of $5M/10%, reduction of funding by ≥10%, etc.) absent notification	As authorized; up to 5% of current fiscal year appropriations may be transferred if appropriations committees are notified at least 30 days in advance; transfer may not represent >10% increase to an individual program except as otherwise specified	Frequent; linked to Disaster Relief Fund for domestic disaster and emergency response and recovery
HHS	• As authorized; no notification below threshold of lesser of $1M or 10% of an account; notification of reprogramming actions above this threshold required • Notification required above threshold of $500K if reprogramming decreases appropriated funding by > 10% or substantially affects program personnel or operations	As authorized; Secretary's One-Percent Transfer General Provision; allows transfer of up to 1% from any account into another account, not to exceed up to 3% of funds previously in account, maximum transfer amount of ~$900M	Frequent; linked to public health crises, hurricane relief, and refugee resettlement support
NASA	• As authorized; reprogramming documents must be submitted if a budget account changes by $500K • Within the Exploration Systems and Space Operations account, no more than 10% of funds for Explorations Systems may be reprogrammed for Space Operations and vice versa	As authorized; transfers for select purposes authorized by 51 U.S.C. § 20143	Rare
ODNI	• As authorized; Director of National Intelligence (DNI) may reprogram funds within the NIP with the approval of the OMB Director and in consultation with affected agencies • Notification to Congress within 30 days for reprogramming actions >$10M or 5% when funds transferred in or out of NIP or between appropriation accounts • Notification to Congress of reprogramming actions prior to June 30	As authorized; DNI may transfer funds within the NIP with the approval of the OMB Director and in consultation with affected agencies	Detailed funding profiles for NIP and MIP are not publicly available.

Abbreviations

AEP	agency execution plan
AOP	agency operating plan
ASA	Office of the Assistant Secretary for Administration
ASFR	Office of the Assistant Secretary for Financial Resources
ASL	Office of the Assistant Secretary for Legislation
ASPE	Office of the Assistant Secretary for Planning and Evaluation
CAM	control account manager
CAPE	Cost Assessment and Program Evaluation
CBJB	congressional budget justification book
CBP	U.S. Customs and Border Protection
CDC	Centers for Disease Control and Prevention
CECR	construction and environmental compliance and restoration
CFO	chief financial officer
CIA	Central Intelligence Agency
CIG	consolidated intelligence guidance
CISA	Cybersecurity and Infrastructure Security Agency
COCOM	combatant command
COP	congressional operating plan
COVID-19	coronavirus disease 2019
CRIB	Community Resource Investment Board
CWMD	Countering Weapons of Mass Destruction
DEXCOM	Deputies Executive Committee
DHS	U.S. Department of Homeland Security
DMAG	Deputy's Management Action Group
DNI	Director of National Intelligence
DoD	U.S. Department of Defense
DoDD	Department of Defense Directive
EXCOM	Executive Committee
FA	federal assistance
FBI	Federal Bureau of Investigation
FDA	Food and Drug Administration
FEMA	Federal Emergency Management Agency
FY	fiscal year
FYDP	Future Years Defense Program

FYHSP	Future Years Homeland Security Program
GAO	U.S. Government Accountability Office
GPRA	Government Performance and Results Act
HEW	U.S. Department of Health, Education, and Welfare
HHS	U.S. Department of Health and Human Services
IC	U.S. intelligence community
ICD	Intelligence Community Directive
ICE	U.S. Immigration and Customs Enforcement
ICS	Intelligence Community Standard
IPG	integrated planning guidance
IPPBE	Intelligence Planning, Programming, Budgeting, and Evaluation
IRIS	Integrated Resources Information System
IRTPA	Intelligence Reform and Terrorism Prevention Act of 2004
IT	information technology
JCIDS	Joint Capabilities Integration and Development System
LHHS	Departments of Labor, Health and Human Services, and Education, and Related Agencies
MIL	major issues list
MIP	Military Intelligence Program
MIS	major issue study
NASA	National Aeronautics and Space Administration
NDAA	National Defense Authorization Act
NDS	National Defense Strategy
NEF	Non-Recurring Expenses Fund
NIH	National Institutes of Health
NIM	national intelligence manager
NIP	National Intelligence Program
NPD	NASA Policy Directive
NPR	NASA Procedural Requirement
O&M	operation and maintenance
O&S	operations and support
OB	Office of Budget
OCFO	Office of the Chief Financial Officer
ODNI	Office of the Director of National Intelligence
OIG	Office of Inspector General
OMB	Office of Management and Budget

OPDIV	operating division
OSD	Office of the Secretary of Defense
P&S	Policy and Strategy Office
PA&E	Office of Program Analysis and Evaluation
PC&I	procurement, construction, and improvements
PDDNI	Principal Deputy Director of National Intelligence
PPAs	programs, projects, and activities
PPBE	Planning, Programming, Budgeting, and Execution
PPBS	Planning, Programming, and Budgeting System
QHSR	Quadrennial Homeland Security Review
R&D	research and development
RAP	resource allocation plan
RAD	resource allocation decision
RC&E	Requirements, Cost, and Effectiveness Division
RDT&E	research, development, test, and evaluation
RPG	resource planning guidance
SAMHSA	Substance Abuse and Mental Health Services Administration
SBC	Secretary's Budget Council
SER	strategic evaluation report
SID	Strategic Investment Division
SIG	Strategic Investment Group
SIPRI	Stockholm International Peace Research Institute
SLC	Senior Leaders Council
SPG	strategic programming guidance
SRA	Systems and Resource Analysis Office
SSMS	safety, security, and mission services
STEM	science, technology, engineering, and mathematics
TSA	Transportation Security Administration
UFMS	Unified Financial Management System
USCG	U.S. Coast Guard

References

Brodie, Bernard, *Strategy in the Missile Age*, RAND Corporation, CB-137-1, 1959. As of April 21, 2023:
https://www.rand.org/pubs/commercial_books/CB137-1.html

CDC—*See* Centers for Disease Control and Prevention.

Centers for Disease Control and Prevention, Financial Management Office, "Financial Management Course," presentation, undated.

Clapper, Jim, and Trey Brown, "A DNI's Overview: Reflection on Integration in the Intelligence Community," *Studies in Intelligence* (extracts), Vol. 65, No. 3, September 2021.

Congressional Research Service, *A Defense Budget Primer*, RL30002, December 9, 1998.

Defense Contract Management Agency Manual 4503-02, *Programming*, U.S. Department of Defense, February 21, 2020.

Department of Defense Directive 5105.84, *Director of Cost Assessment and Program Evaluation*, Office of the Chief Management Officer of the Department of Defense, August 14, 2020.

Department of Defense Directive 5205.12, *Military Intelligence Program*, change 2, October 1, 2020.

Department of Defense Directive 7045.14, *The Planning, Programming, Budgeting, and Execution (PPBE) Process*, U.S. Department of Defense, August 29, 2017.

Department of Homeland Security Directive 101-01, *Planning, Programming, Budgeting, and Execution*, rev. 1, U.S. Department of Homeland Security, June 4, 2019.

Department of Homeland Security Instruction 069-03-001, *Program, Policy, and Organizational Evaluations*, U.S. Department of Homeland Security, February 16, 2021.

Department of Homeland Security Instruction 101-01-001, *Planning, Programming, Budgeting, and Execution*, U.S. Department of Homeland Security, June 11, 2019.

DeVine, Michael E., *Defense Primer: Budgeting for National and Defense Intelligence*, Congressional Research Service, IF10524, November 17, 2022.

DHS—*See* U.S. Department of Homeland Security.

DHS Directive—*See* Department of Homeland Security Directive.

DHS Instruction—*See* Department of Homeland Security Instruction.

DHS OIG—*See* U.S. Department of Homeland Security, Office of Inspector General.

Directive-Type Memorandum 09-027, Implementation of the Weapon Systems Acquisition Reform Act of 2009, Office of the Under Secretary of Defense for Acquisition, Technology and Logistics, U.S. Department of Defense, December 4, 2009, incorporating change 1, October 21, 2010.

Director of Central Intelligence Directive 3/3, *Community Management Staff*, Central Intelligence Agency, June 12, 1995.

DoD—*See* U.S. Department of Defense.

DoDD—*See* Department of Defense Directive.

Enthoven, Alain C., and K. Wayne Smith, *How Much Is Enough? Shaping the Defense Program, 1961–1969*, RAND Corporation, CB-403, 1971. As of April 21, 2023:
https://www.rand.org/pubs/commercial_books/CB403.html

Executive Order 12333, *United States Intelligence Activities*, Executive Office of the President, December 4, 1981.

GAO—*See* U.S. Government Accountability Office.

Greenwalt, William, and Dan Patt, *Competing in Time: Ensuring Capability Advantage and Mission Success Through Adaptable Resource Allocation*, Hudson Institute, February 2021.

Harlow, Robert L., "On the Decline and Possible Fall of PPBS," *Public Finance Quarterly*, Vol. 1, No. 2, April 1973.

HHS—*See* U.S. Department of Health and Human Services.

Hitch, Charles J., and Roland N. McKean, *The Economics of Defense in the Nuclear Age*, RAND Corporation, R-346, 1960. As of April 21, 2023:
https://www.rand.org/pubs/reports/R346.html

ICD—*See* Intelligence Community Directive.

ICS—*See* Intelligence Community Standard.

Intelligence Community Directive 104, *National Intelligence Program (NIP) Budget Formulation and Justification, Execution, and Performance Evaluation*, Office of the Director of National Intelligence, April 30, 2013.

Intelligence Community Directive 116, *Intelligence Planning, Programming, Budgeting, and Evaluation System*, Office of the Director of National Intelligence, September 14, 2011.

Intelligence Community Directive 801, *Acquisition*, Office of the Director of National Intelligence, August 16, 2009.

Intelligence Community Directive 900, *Integrated Mission Management*, Office of the Director of National Intelligence, May 6, 2013.

Intelligence Community Policy Guidance 801.1 Technical Amendment, *Acquisition*, Office of the Director of National Intelligence, January 29, 2015.

Intelligence Community Standard 104-01, *Engagement with Congress on National Intelligence Program (NIP) Budgeting Activities*, Office of the Director of National Intelligence, June 27, 2018.

Intelligence Community Standard 104-02, *National Intelligence Program (NIP) Procedures for Reprogramming and Transfer Actions*, Office of the Director of National Intelligence, June 26, 2018.

Jablonsky, Stephen F., and Mark W. Dirsmith, "The Pattern of PPB Rejection: Something About Organizations, Something About PPB," *Accounting, Organizations and Society*, Vol. 3, Nos. 3–4, 1978.

Johnson, Jeh, Secretary of Homeland Security, "Strengthening Departmental Unity of Effort," memorandum for DHS leadership, April 22, 2014.

Kringen, John A., "Rethinking the Concept of Global Coverage in the US Intelligence Community," *Studies in Intelligence*, Vol. 59, No. 3, September 2015.

Long, Dave, "CBP's Eyes in the Sky," *Frontline Magazine*, U.S. Customs and Border Protection, last updated April 11, 2016.

Lowenthal, Mark M., *Intelligence: From Secrets to Policy*, 8th ed., CQ Press, 2019.

Luckey, David, David Stebbins, Sarah W. Denton, Elena Wicker, Stephanie Anne Pillion, and Alice Shih, *National Geospatial-Intelligence Agency Resources: Financial Management Programming Evaluation*, RAND Corporation, RR-A659-1, 2022. As of March 13, 2023: https://www.rand.org/pubs/research_reports/RRA659-1.html

McGarry, Brendan W., *Defense Primer: Planning, Programming, Budgeting and Execution (PPBE) Process*, Congressional Research Service, IF10429, January 27, 2020.

McGarry, Brendan W., *DOD Planning, Programming, Budgeting, and Execution: Overview and Selected Issues for Congress*, Congressional Research Service, R47178, July 11, 2022.

McKernan, Megan, Stephanie Young, Andrew Dowse, James Black, Devon Hill, Benjamin J. Sacks, Austin Wyatt, Nicolas Jouan, Yuliya Shokh, Jade Yeung, Raphael S. Cohen, John P. Godges, Heidi Peters, and Lauren Skrabala, *Planning, Programming, Budgeting, and Execution in Comparative Organizations:* Vol. 2, *Case Studies of Selected Allied and Partner Nations*, RAND Corporation, RR-A2195-2, 2024. As of January 12, 2024: https://www.rand.org/pubs/research_reports/RRA2195-2.html

McKernan, Megan, Stephanie Young, Timothy R. Heath, Dara Massicot, Andrew Dowse, Devon Hill, James Black, Ryan Consaul, Michael Simpson, Sarah W. Denton, Anthony Vassalo, Ivana Ke, Mark Stalczynski, Benjamin J. Sacks, Austin Wyatt, Jade Yeung, Nicolas Jouan, Yuliya Shokh, William Shelton, Raphael S. Cohen, John P. Godges, Heidi Peters, and Lauren Skrabala, *Planning, Programming, Budgeting, and Execution in Comparative Organizations:* Vol. 4, *Executive Summary*, RAND Corporation, RR-A2195-4, 2024. As of January 12, 2024: https://www.rand.org/pubs/research_reports/RRA2195-4.html

McKernan, Megan, Stephanie Young, Timothy R. Heath, Dara Massicot, Mark Stalczynski, Ivana Ke, Raphael S. Cohen, John P. Godges, Heidi Peters, and Lauren Skrabala, *Planning, Programming, Budgeting, and Execution in Comparative Organizations:* Vol. 1, *Case Studies of China and Russia*, RAND Corporation, RR-A2195-1, 2024. As of January 12, 2024: https://www.rand.org/pubs/research_reports/RRA2195-1.html

Mirabello, Robert A., "Guide to the Study of Intelligence: Budget and Resource Management," *The Intelligencer: Journal of U.S. Intelligence Studies*, Vol. 20, No. 2, Fall/Winter 2013.

Mitchell, Alison, *Medicaid Financing and Expenditures*, Congressional Research Service, R42640, November 10, 2020.

Morgan, Daniel, *NASA Appropriations and Authorizations: A Fact Sheet*, Congressional Research Service, July 2, 2021.

Mrdeza, Michelle, and Kenneth Gold, "Reprogramming Funds: Understanding the Appropriators' Perspective," Government Affairs Institute at Georgetown University, undated.

NASA—*See* National Aeronautics and Space Administration.

NASA OCFO—*See* National Aeronautics and Space Administration, Office of the Chief Financial Officer.

NASA OIG—*See* National Aeronautics and Space Administration, Office of Inspector General.

NASA Science, "Most Recent Decadal Studies," webpage, March 9, 2023. As of March 10, 2023: https://science.nasa.gov/about-us/science-strategy/decadal-surveys

National Aeronautics and Space Administration Policy Directive 1000.0C, *NASA Governance and Strategic Management Handbook*, Office of the Associate Administrator, National Aeronautics and Space Administration, January 29, 2020.

NASA Procedural Requirement 9420.1A, *Budget Formulation*, Office of the Chief Financial Officer, National Aeronautics and Space Administration, incorporating change 1, September 15, 2021.

National Aeronautics and Space Administration Procedural Requirement 9470.1, *Budget Execution*, Office of the Chief Financial Officer, National Aeronautics and Space Administration, December 24, 2008.

National Aeronautics and Space Administration, "NASA Establishes Office of Program Analysis and Evaluation," press release, June 20, 2005.

National Aeronautics and Space Administration, *2022 Strategic Plan*, 2022a.

National Aeronautics and Space Administration, *FY 2022 Agency Financial Report*, Section 3, "Summary of Financial Statement Audit and Management Assurances," 2022b.

National Aeronautics and Space Administration, *FY 2023 President's Budget Request Summary*, 2022c.

National Aeronautics and Space Administration, organizational chart, February 25, 2022d.

National Aeronautics and Space Administration, "Missions," webpage, last updated April 15, 2022e. As of October 13, 2023:
https://www.nasa.gov/nasa-missions/

National Aeronautics and Space Administration, *Economic Impact Report*, October 2022f.

National Aeronautics and Space Administration, "NASA Receives 12th Successive 'Clean' Financial Audit Rating," press release, November 15, 2022g.

National Aeronautics and Space Administration, "About NASA," webpage, last updated January 27, 2023a. As of March 10, 2023:
https://www.nasa.gov/about

National Aeronautics and Space Administration, "Previous Years' Budget Requests," webpage, last updated February 23, 2023b. As of October 13, 2023:
https://www.nasa.gov/previous-years-budget-requests/

National Aeronautics and Space Administration, Office of Inspector General, *Integrated Financial Management Program Budget Formulation Module*, Audit Report IG-04-017, March 30, 2004.

National Aeronautics and Space Administration, Office of the Chief Financial Officer, "Congressional Committees," webpage, last updated March 25, 2019. As of March 10, 2023:
https://www.nasa.gov/offices/ocfo/appropriations/congressional_committees

National Institutes of Health, Office of Management and Assessment, "Budget Execution," Chapter 1920, *NIH Policy Manual*, March 10, 2020.

NIH—*See* National Institutes of Health.

NPD—*See* National Aeronautics and Space Administration Policy Directive.

NPR—*See* National Aeronautics and Space Administration Procedural Requirement.

ODNI—*See* Office of the Director of National Intelligence.

Office of Management and Budget, "Historical Tables," White House, undated. As of January 11, 2023:
https://www.whitehouse.gov/omb/budget/historical-tables

Office of Management and Budget Circular No. A-11, *Preparation, Submission, and Execution of the Budget*, Executive Office of the President, August 2022.

Office of the Director of National Intelligence, "Chief Financial Officer," webpage, undated-a. As of March 13, 2023:
https://www.dni.gov/index.php/features/219-about/organization/chief-financial-officer

Office of the Director of National Intelligence, "IC Planning Process," undated-b.

Office of the Director of National Intelligence, "Intelligence Community Resource Process Guide," undated-c.

Office of the Director of National Intelligence, "Mission, Vision & Values," webpage, undated-d. As of March 13, 2023:
https://www.dni.gov/index.php/who-we-are/mission-vision

Office of the Director of National Intelligence, "Policy & Capabilities–Who We Are," webpage, undated-e. As of March 13, 2023:
https://www.dni.gov/index.php/who-we-are/organizations/policy-capabilities/policy-capabilities-who-we-are

Office of the Director of National Intelligence, "U.S. Intelligence Community Budget," webpage, undated-f. As of January 30, 2023:
https://www.dni.gov/index.php/what-we-do/ic-budget

Office of the Director of National Intelligence, "What We Do," webpage, undated-g. As of March 13, 2023:
https://www.dni.gov/index.php/what-we-do

Office of the Director of National Intelligence, *U.S. National Intelligence: An Overview*, 2011a.

Office of the Director of National Intelligence, "NIP-MIP Rules of the Road," May 2011b.

Office of the Director of National Intelligence, "ODNI Fact Sheet," October 2011c.

Office of the Director of National Intelligence, "ODNI Factsheet," February 24, 2017.

Office of the Director of National Intelligence, "ODNI Completes Transformation Effort," press release, July 25, 2018.

Office of the Director of National Intelligence, *National Intelligence Strategy of the United States of America*, 2019.

Office of the Director of National Intelligence, "ODNI Implements Organizational Realignment to Increase Effectiveness," press release, August 17, 2020.

Office of the Director of National Intelligence, organizational chart, October 20, 2022.

Office of the Secretary of Defense, Cost Assessment and Program Evaluation, homepage, undated. As of March 10, 2023:
https://www.cape.osd.mil

Office of the Under Secretary of Defense (Comptroller), "Summary of Reprogramming Requirements, Effective for FY 2021 Appropriation," January 6, 2021.

Office of the Under Secretary of Defense for Acquisition and Sustainment, "Budget Activity (BA) 'BA-08': Software and Digital Technology Pilot Program Frequently Asked Questions," version 1.0, U.S. Department of Defense, September 28, 2020.

OMB—*See* Office of Management and Budget.

OSD—*See* Office of the Secretary of Defense.

Painter, William L., *DHS Budget v. DHS Appropriations: Fact Sheet*, Congressional Research Service, R44052, April 17, 2019.

Painter, William L., *Department of Homeland Security Appropriations: FY2021*, Congressional Research Service, R46802, May 24, 2021.

Painter, William L., *Department of Homeland Security Appropriations: FY2022*, Congressional Research Service, R47005, March 24, 2022.

Painter, William L., Michael E. DeVine, Bart Elias, Kristin Finklea, John Frittelli, Jill C. Gallagher, Frank Gottron, Diane P. Horn, Chris Jaikaran, Lennard G. Kruger, et al., *Selected Homeland Security Issues in the 116th Congress*, Congressional Research Service, R45701, November 26, 2019.

Painter, William L. and Audrey Singer, *DHS Border Barrier Funding*, Congressional Research Service, R45888, January 29, 2020.

Public Law 85-568, National Aeronautics and Space Act of 1958, July 29, 1958.

Public Law 103-62, Government Performance and Results Act of 1993, August 3, 1993.

Public Law 108-458, Intelligence Reform and Terrorism Prevention Act of 2004, December 17, 2004.

Public Law 111-23, Weapon Systems Acquisition Reform Act of 2009, May 22, 2009.

Public Law 111-352, GPRA Modernization Act of 2010, January 4, 2011.

Public Law 115-414, Good Accounting Obligation in Government Act of 2019, January 3, 2019.

Public Law 115-435, Foundations for Evidence-Based Policymaking Act of 2018, January 14, 2019.

Public Law 116-117, Payment Integrity Information Act of 2019, March 2, 2020.

Public Law 116-123, Coronavirus Preparedness and Response Supplemental Appropriations Act, 2020, March 6, 2020.

Public Law 117-81, National Defense Authorization Act for Fiscal Year 2022, December 27, 2021.

Public Law 117-103, Consolidated Appropriations Act, 2022, March 15, 2022.

Radin, Beryl A., *Managing Decentralized Departments: The Case of the U.S. Department of Health and Human Services*, PricewaterhouseCoopers Endowment for the Business of Government, October 1999.

Rappaport, Carl S., "Program Budgeting and PPBS in the Federal Government," paper presented at the 48th Annual Meeting of the Committee on Program Budgeting, 1969.

Rivlin, Alice M., "The Planning, Programming, and Budgeting System in the Department of Health, Education, and Welfare: Some Lessons from Experience," in U.S. House of Representatives, Joint Economic Committee, Subcommittee on Economy in Government, *The Analysis and Evaluation of Public Expenditures: The PPB System, A Compendium of Papers Submitted to the Subcommittee on Economy in Government of the Joint Economic Committee*, Vol. 3, Part V, Section C, U.S. Government Printing Office, 1969, pp. 909–922.

Rosenwasser, Jon, "Intelligence Integration: A Congressional Oversight Perspective," *Studies in Intelligence* (extracts), Vol. 65, No. 3, September 2021.

Schick, Allen, "A Death in the Bureaucracy: The Demise of Federal PPB," *Public Administration Review*, Vol. 33, No. 2, March–April 1973.

Section 809 Panel, *Report of the Advisory Panel on Streamlining and Codifying Acquisition Regulations*, Vol. 2 of 3, June 2018.

Sekar, Kavya, *COVID-19 Supplemental Appropriations for the Department of Health and Human Services (HHS), 2022: In Brief*, Congressional Research Service, R47232, October 7, 2022.

SIPRI—*See* Stockholm International Peace Research Institute.

Speciale, Stephen, and Wayne B. Sullivan II, "DoD Financial Management—More Money, More Problems," Defense Acquisition University, September 1, 2019.

Stockholm International Peace Research Institute, "SIPRI Military Expenditure Database," homepage, undated. As of March 17, 2023:
https://milex.sipri.org/sipri

Tollestrup, Jessica, Karen E. Lynch, and Ada S. Cornell, *Department of Health and Human Services: FY2023 Budget Request*, Congressional Research Service, R47122, May 31, 2022.

U.S. Code, Title 10, Section 3131, Availability of Appropriations.

U.S. Code, Title 50, Section 3024, Responsibilities and Authorities of the Director of National Intelligence.

U.S. Code, Title 50, Section 3094, Funding of Intelligence Activities.

U.S. Code, Title 50, Section 3108, Auditability of Certain Elements of the Intelligence Community.

U.S. Code, Title 51, Section 20143, Full Cost Appropriations Account Structure.

U.S. Constitution, Article II, Section 3.37, Impounding Appropriated Funds.

U.S. Department of Defense, "About," webpage, undated. As of January 20, 2023:
https://www.defense.gov/About

U.S. Department of Defense, *2022 National Defense Strategy of the United States of America*, 2022.

U.S. Department of Health and Human Services, "Fiscal Year 2017 Nonrecurring Expenses Fund: Justification of Estimates for Appropriations Committee," undated.

U.S. Department of Health and Human Services, "Office of the Assistant Secretary for Financial Resources Functional Statement," webpage, last reviewed May 2, 2017. As of March 6, 2023:
https://www.hhs.gov/about/agencies/asfr/functional-statement/index.html

U.S. Department of Health and Human Services, "Strategic Plan FY 2022–2026," webpage, last reviewed March 28, 2022a. As of March 16, 2023:
https://www.hhs.gov/about/strategic-plan/2022-2026/index.html

U.S. Department of Health and Human Services, "FY 2023 HHS Contingency Staffing Plan for Operations in the Absence of Enacted Annual Appropriations," webpage, last reviewed October 5, 2022b. As of April 4, 2023:
https://www.hhs.gov/about/budget/fy-2023-hhs-contingency-staffing-plan/index.html

U.S. Department of Health and Human Services, "HHS Historical Highlights," webpage, last reviewed March 14, 2023a. As of March 16, 2023:
https://www.hhs.gov/about/historical-highlights/index.html

U.S. Department of Health and Human Services, "HHS Organizational Charts: Office of Secretary and Divisions," webpage, last reviewed August 17, 2023b. As of August 21, 2023:
https://www.hhs.gov/about/agencies/orgchart/index.html

U.S. Department of Homeland Security, "Department of Homeland Security's Strategic Plan for Fiscal Years 2020–2024," webpage, undated-a. As of February 6, 2023:
https://www.dhs.gov/publication/department-homeland-securitys-strategic-plan-fiscal-years-2020-2024

U.S. Department of Homeland Security, *U.S. Department of Homeland Security FY 2020–2022 Annual Performance Report*, undated-b.

U.S. Department of Homeland Security, *Financial Management Policy Manual*, September 8, 2020, Not available to the general public.

U.S. Department of Homeland Security, "History," webpage, last updated April 26, 2022a. As of December 30, 2022:
https://www.dhs.gov/history

U.S. Department of Homeland Security, *U.S. Department of Homeland Security Agency Financial Report, FY 2022*, November 15, 2022b.

U.S. Department of Homeland Security, Office of Inspector General, *DHS' H-60 Helicopter Programs*, OIG-13-89, revised May 2013.

U.S. Department of Homeland Security, Office of Inspector General, *Independent Auditors' Report on the Department of Homeland Security's Consolidated Financial Statements for FYs 2022 and 2021 and Internal Control over Financial Reporting*, OIG-23-02, November 15, 2022.

U.S. Department of Homeland Security, Office of the Chief Financial Officer, *A Common Appropriations Structure for DHS: FY 2016 Crosswalk*, Addendum to the Fiscal Year 2016 President's Budget, February 2, 2015.

U.S. General Accounting Office, *The Budget and Accounting Act*, 1966.

U.S. General Accounting Office, *Management of HHS: Using the Office of the Secretary to Enhance Departmental Effectiveness*, GAO-HRD-90-54, February 9, 1990.

U.S. General Accounting Office, *Managing for Results: Efforts to Strengthen the Link Between Resources and Results at the Administration for Children and Families*, GAO-03-9, December 10, 2002.

U.S. Government Accountability Office, *A Glossary of Terms Used in the Federal Budget Process*, GAO-05-734SP, September 2005.

U.S. Government Accountability Office, *Quadrennial Homeland Security Review: Improved Risk Analysis and Stakeholder Consultations Could Enhance Future Reviews*, GAO-16-371, April 2016a.

U.S. Government Accountability Office, *DHS Management: Enhanced Oversight Could Better Ensure Programs Receiving Fees and Other Collections Use Funds Efficiently*, GAO-16-443, July 21, 2016b.

U.S. Government Accountability Office, *Homeland Security: Clearer Roles and Responsibilities for the Office of Strategy, Policy, and Plans and Workforce Planning Would Enhance Its Effectiveness*, GAO-18-590, September 2018.

U.S. Government Accountability Office, *Department of Homeland Security: Progress Made Strengthening Management Functions, but Work Remains*, GAO-21-105418, September 30, 2021.

U.S. Government Accountability Office, *DHS Annual Assessment: Most Acquisition Programs Are Meeting Goals Even with Some Management Issues and COVID-19 Delays*, GAO-22-104684, March 8, 2022.

U.S. House of Representatives, Committee on Appropriations, *Department of Homeland Security Appropriations Bill, 2019*, House Report 115-948, U.S. Government Publishing Office, September 12, 2018.

West, William F., *Program Budgeting and the Performance Movement: The Elusive Quest for Efficiency in Government*, Georgetown University Press, 2011.

Williams, McCoy, "Department of Homeland Security: Financial Management Challenges," testimony before the Subcommittee on Financial Management, the Budget, and International Security, Committee on Governmental Affairs, U.S. Senate, GAO-04-945T, U.S. General Accounting Office, July 8, 2004.

RAND NATIONAL DEFENSE RESEARCH INSTITUTE

The U.S. Department of Defense (DoD) Planning, Programming, Budgeting, and Execution (PPBE) System is a key enabler for DoD to fulfill its mission. But in light of a dynamic threat environment, increasingly capable adversaries, and rapid technological changes, there has been increasing concern that DoD's resource planning processes are too slow and inflexible to meet warfighter needs. As a result, Congress mandated the formation of a legislative commission to (1) examine the effectiveness of the PPBE process and adjacent DoD practices, particularly with respect to defense modernization; (2) consider potential alternatives to these processes and practices to maximize DoD's ability to respond in a timely manner to current and future threats; and (3) make legislative and policy recommendations to improve such processes and practices for the purposes of fielding the operational capabilities necessary to outpace near-peer competitors, providing data and analytical insight, and supporting an integrated budget that is aligned with strategic defense objectives.

The Commission on PPBE Reform asked the RAND Corporation to provide an independent analysis of PPBE-like functions in selected countries and other non-DoD federal agencies. This report, part of a four-volume set, presents case studies of PPBE functions in the U.S. Department of Homeland Security, the U.S. Department of Health and Human Services, the National Aeronautics and Space Administration, and the Office of the Director of National Intelligence to provide insights for improving DoD's PPBE processes.

$39.00

ISBN-10 1-9774-1238-6
ISBN-13 978-1-9774-1238-6

53900

9 781977 412386

www.rand.org